Research and Education 2018

Transportation Research Record: Journal of the Transportation Research Board

Volume 2672, Issue 13

ISBN: 978-1-5108-7822-8

Printed from e-media with permission by:

Curran Associates, Inc.
57 Morehouse Lane
Red Hook, NY 12571

Some format issues inherent in the e-media version may also appear in this print version.

Published by Sage Publications USA

ISBN (Print) 978-1-5108-7822-8
ISBN 2018 Printed Set (All Issues) 978-1-5108-7822-8

Printed by Curran Associates, Inc. (2019)

Additional copies of this publication are available from:

Curran Associates, Inc.
57 Morehouse Lane
Red Hook, NY 12571 USA
Phone: 845-758-0400
Fax: 845-758-2633
Email: curran@proceedings.com
Web: www.proceedings.com

Volume 2672 Number 13 December 2018

TRR

JOURNAL OF THE TRANSPORTATION RESEARCH BOARD

Contents

Research and Education

Articles

TRB

TRANSPORTATION RESEARCH BOARD

CORE SPONSORS

Transportation Departments of the 50 States and the District of Columbia

Federal Government

U.S. Department of Transportation
- Federal Aviation Administration (FAA)
- Federal Highway Administration
- Federal Motor Carrier Safety Administration (FMCSA)
- Federal Railroad Administration (FRA)
- Federal Transit Administration
- National Highway Traffic Safety Administration (NHTSA)
- Office of the Assistant Secretary of Transportation for Research and Technology (OST-R)

U.S. Air Force Civil Engineer Center (AFCEC)
U.S. Coast Guard (USCG)
U.S. Department of Energy
U.S. Department of the Interior (DOI)
U.S. Environmental Protection Agency (EPA)

Nongovernmental Organizations

American Public Transportation Association (APTA)
Association of American Railroads (AAR)
California Air Resources Board

The National Academies of

SCIENCES · ENGINEERING · MEDICINE

The **National Academy of Sciences** was established in 1863 by an Act of Congress, signed by President Lincoln, as a private, nongovernmental institution to advise the nation on issues related to science and technology. Members are elected by their peers for outstanding contributions to research. Dr. Marcia McNutt is president.

The **National Academy of Engineering** was established in 1964 under the charter of the National Academy of Sciences to bring the practices of engineering to advising the nation. Members are elected by their peers for extraordinary contributions to engineering. Dr. C. D. Mote, Jr., is president.

The **National Academy of Medicine** (formerly the Institute of Medicine) was established in 1970 under the charter of the National Academy of Sciences to advise the nation on medical and health issues. Members are elected by their peers for distinguished contributions to medicine and health. Dr. Victor J. Dzau is president.

The three Academies work together as the National Academies of Sciences, Engineering, and Medicine to provide independent, objective analysis and advice to the nation and conduct other activities to solve complex problems and inform public policy decisions. The Academies also encourage education and research, recognize outstanding contributions to knowledge, and increase public understanding in matters of science, engineering, and medicine.

Learn more about the National Academies of Sciences, Engineering, and Medicine at **www.national-academies.org.**

The **Transportation Research Board** is one of seven major programs of the National Academies of Sciences, Engineering, and Medicine. The mission of the Transportation Research Board is to increase the benefits that transportation contributes to society by providing leadership in transportation innovation and progress through research and information exchange, conducted within a setting that is objective, interdisciplinary, and multimodal. The Board's varied activities annually engage about 7,000 engineers, scientists, and other transportation researchers and practitioners from the public and private sectors and academia, all of whom contribute their expertise in the public interest. The program is supported by state transportation departments, federal agencies including the component administrations of the U.S. Department of Transportation, and other organizations and individuals interested in the development of transportation.

Learn more about the Transportation Research Board at **www.TRB.org**.

TRR
JOURNAL OF THE TRANSPORTATION RESEARCH BOARD

Transportation Research Record
2018, Vol. 2672(13) 1–9

DOI: 10.1177/0361198118757968
journals.sagepub.com/home/trr
SAGE

Decision Support for Civil Engineering Students: Analysis of Alumni Career Paths

Susan L. Hotle[1] and Bryan J. Katz[1]

Abstract

Undergraduate students in engineering face many important decisions in the final 2 years of their degree program. These decisions can have an impact on long-term career choices, such as specialization area, career role of interest, and whether to apply to graduate school. Unfortunately, uninformed decisions can lead to missed opportunities, as well as the student potentially leaving science, technology, engineering, and mathematics education due to choosing a specialization that is not well aligned with their interests. This survey-based study assists students by analyzing the personality types, demographics, and career paths of 567 alumni that have earned an undergraduate degree in civil and environmental engineering (CEE) and are no longer enrolled in a university. Study findings include the fact that certain demographics, personality types, and job preferences are significant predictors of the final outcome of an alumni's career when choosing between the different technical areas within CEE and professional roles. Family history of having an engineer in the immediate family did not prove to be a significant factor in these decisions. In addition, little significance was found between the data captured in the survey of whether or not someone would go on to earn a graduate degree in CEE. Given where significant relationships were found, it is recommended that future studies focus on testing additional personality types (e.g., is enthusiastic) and job traits (e.g., likes a desk job) to provide even greater distinctions between the technical areas and roles.

To remain internationally competitive and account for evolving workforce needs, there has been a focus on promoting science, technology, engineering, and mathematics (STEM) education and making it more accessible in the United States, even at an early age (*1*). Student recruitment for these fields has been of upmost focus in the education system, as the Bureau of Labor Statistics found a 10.5% growth in the number of STEM occupations from 2009–2015, compared with 5.2% growth of non-STEM occupations (*2*). Professionals in STEM fields are also expected to be highly competitive for non-STEM openings due to their critical-thinking, analysis, and problem-solving skills, with recent findings showing a "… deeper understanding of STEM is tied to a more productive overall workforce—not just STEM industries" (*3*).

To evaluate the success of STEM education outreach, current literature has predominately evaluated factors that affect recruitment (*4*), persistence (*5*), and attrition (*6*). These studies analyzed not only the rates, but also how equitable the rates are among different groups; for example, with respect to gender, race, or income. Although these findings help realign future investment strategies from a government standpoint, these findings are not as useful from a student's perspective. That is, these results become diagnostic tests, informing students about how more or less likely they are to succeed in STEM, given characteristics about themselves. To minimize attrition rates, students arguably would benefit more from a prescriptive study. That is, given a student's history and characteristics, where in engineering would they be predicted to have a career best aligned with their interests? In which STEM field do alumni most similar to them in the workforce end up?

A prescriptive study benefits students in several ways. First, it could help students dealing with career indecisiveness. As stated in Cantrell and Ewing-Taylor (*7*), "when adolescents explore their own traits along with the characteristics of career options they are considering, they tend to avoid problems later on such as career indecision or prematurely making a poorly founded career decision." This indecisiveness can be especially costly if it occurs later on in a college student's degree program (e.g., the last 2 years) when they face questions about which area to specialize in within the major, and whether or not to pursue a graduate degree before entering the workforce. Second, this type of study would recommend an alternative STEM field for students to try before changing to a non-STEM major, therefore potentially reducing the attrition rate. Third, a study based on alumni that have

[1]Department of Civil and Environmental Engineering, Virginia Polytechnic Institute and State University, Blacksburg, VA

Corresponding Author:
Address correspondence to Susan L. Hotle: shotle3@vt.edu

worked for several years in the field could help indicate which area a student may have a long-term career with a reduced chance of burnout. For example, previous studies in the medical field have found that "although the causes of burnout are multifactorial, individual characteristics play an important role" (*8*).

The purpose of this study is to provide current undergraduate students—specifically in civil and environmental engineering (CEE)—with decision support during their undergraduate careers. CEE degree programs typically encourage students to specialize through required technical electives in their junior and senior years. These specialization decisions for CEE majors are complex, as it is known to be a very diverse field that trains students in anything from designing concrete with a specific strength to forecasting multimodal demand in 10 years. Therefore, this study analyzes career paths of CEE alumni to provide guidance to current students in the final years of pursuing their undergraduate degrees in the field. This study was motivated by evident indecisiveness of students enrolled in 3000- and 4000-level courses taught by the authors.

Methodology

Using the online survey software Qualtrics®, CEE alumni were questioned on their demographics, family and educational background, and personality traits. The survey solicitation was distributed via email to Virginia Tech's CEE alumni as well as posted to the group's social media page. The alumni could forward the survey link to other professionals in their network that also met the study's two inclusion criteria: (1) must have earned an undergraduate CEE degree; (2) cannot currently be enrolled at a university. The second criterion prevents current CEE graduate students from participating, as the goal of this study was to analyze the professional careers and permanent placement of those with an undergraduate degree in CEE. In addition, the authors reached out to contacts at private companies to specifically ask that the survey be sent to all graduates with CEE degrees, not just from Virginia Tech.

Table 1 lists the information collected from each respondent. The survey first asked alumni questions that current undergraduate CEE students could answer about themselves. These predictor variables included demographic information, personality type tested through Likert-scale questions, and a description of the circumstances at the beginning of their undergraduate degree program. For example: At what age did the alumni begin their undergraduate CEE degree? Were either their mother, father, or any immediate family member working in the engineering field at the time of their admittance to the CEE degree program? Alumni were then asked about their current careers to provide the dependent variables, such as the technical area in CEE they currently work in, what job role they have, and whether they completed a graduate-level CEE degree after completing their undergraduate studies. Alumni were also asked questions about their preferred career to indicate whether their current career matches their interests.

Data

The survey was fully completed by a total of 567 CEE alumni. Most responses were received from Virginia Tech alumni (52.2%) owing, to the alumni association solicitation channel. Responses from other universities were made possible by Virginia Tech alumni forwarding the online survey link to people in their professional network that met the survey criteria. The university with second most frequent responses was Texas A&M, with 3.6% of respondents.

Of the respondents, 68.4% were male. The ethnical affiliations of respondents were 87.8% white, 5.6% Hispanic, 4.2% Asian, and 1.4% black/African American. This means that the study oversampled females and whites, who in 2015 were awarded 24.2% and 66.8% of bachelor's degrees in CEE, respectively (*9*). At the start of their undergraduate CEE degree, 97.5% of respondents were U.S. citizens, 91.7% were between the ages of 17–19 years old, and 68.8% were considered in-state at the university they attended. Most respondents were somewhat new in their careers, with 69.5% beginning their undergraduate degree in CEE in 2000 or later. The highest degree earned in CEE for respondents was 80.4%, 0.2%, 17.5%, and 1.9% for B.S., M.E. (Master's in Engineering with a CEE focus), M.S., and PhD, respectively.

Figure 1 presents the distribution of the careers reported by the alumni. The technical areas were based on the divisions of the American Society of Civil Engineering (*10*). Alumni that responded "other" cited careers in, for example, information technology, tax accounting, legal services, finance, and facility management. Participants could select up to two technical areas; 19% of respondents selected two (i.e., "multiple"). It is important to note that 93.8% of responses agreed that their current job fits their interests and 4.6% were neutral.

Figure 2 outlines the role each of the respondents fills at their current position. Half of respondents were in the practice/design side of CEE, followed by 29% who selected two roles. Most teaching responses were coupled with another role, therefore putting the samples in the multiple category. In addition, these roles were not evenly distributed by the different technical areas, as earning a CEE degree opens up different opportunities in each area. For example, 74% of respondents in construction cited having a management-only role, but management was only 13% of all 567 responses. Similarly, geotechnical, land development, structures, and transportation had increased responses of practice/design.

Table 1. Information Captured in Survey

Category	Variable
Demographics	Ethnicity
	Gender (Female = 0, Male = 1)
Status at start of undergraduate CEE degree	Father engineer (No = 0, Yes = 1)
	Mother engineer (No = 0, Yes = 1)
	Engineer in family (No = 0, Yes = 1)
	Country of citizenship
	In/out state status
	Year
	Age
Personality statements using the Likert scale (0 = Disagree, 1 = Neutral, 2 = Agree)	I like to take risks
	I prefer to work alone over working in groups
	I prefer a desk job over working out in the field
	I tend to procrastinate
	I work well under pressure
	I am enthusiastic
	I am a critical thinker
	I am dependable
	I am anxious
	I am open to new experiences
	I am reserved, quiet
	I am sympathetic, warm
	I am disorganized
	I am calm
Earned CEE degree information	Undergraduate CEE degree university
	Highest CEE degree obtained
Current career	Current technical area (e.g., structures, geotechnical, etc.). Select up to 2.
	Current role (e.g., managing, practice/design, etc.). Select up to 2.
Preferred career	Preferred technical area (e.g., structures, geotechnical, etc.). Select up to 2.
	Preferred role (e.g., managing, practice/design, etc.). Select up to 2.
	I believe my current job fits my interests. Uses Likert scale.

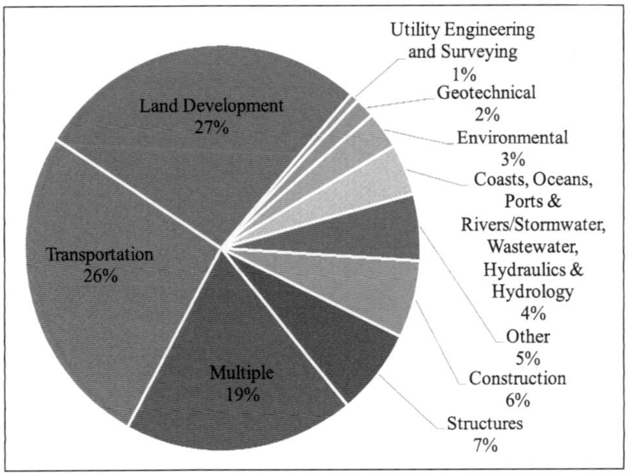

Figure 1. Current technical areas of study participants.

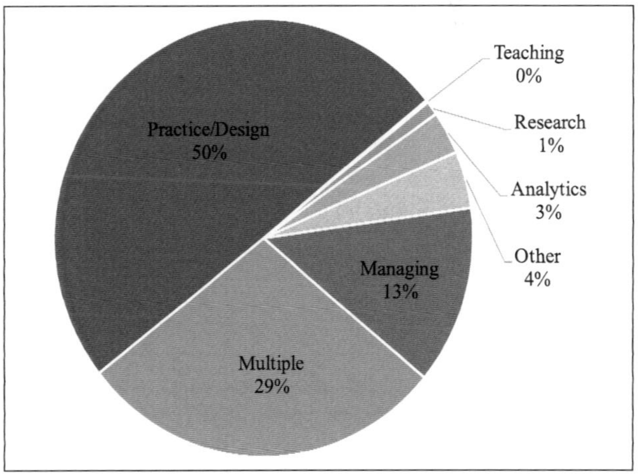

Figure 2. Current roles of study participants.

Results

Table 2 shows the Spearman correlations between 18 of the variables collected in the survey, also indicating those relationships that are at least 95% significant. Spearman correlations were used as many of the variables are Likert-scale responses (i.e., are ordinal) and this method should be used when comparing variables that have "violations of

Table 2. Correlations between Respondent Characteristics

	1	2	3	4	5	6	7	8	9	10	11	12	13	14	15	16	17	18
1. Gender	1																	
2. Father engineer	−0.06	1																
3. Mother engineer	−0.07	0.22*	1															
4. Engineer in family	−0.08	0.82*	0.31*	1														
5. Likes to take risks	0.08	−0.02	−0.00	−0.02	1													
6. Likes to work alone	0.09	0.06	−0.11*	0.03	−0.12*	1												
7. Likes desk job	−0.06	0.02	−0.04	0.03	−0.18*	−0.06	1											
8. Tends to procrastinate	0.07	0.09	0.04	0.09*	−0.03	0.02	−0.00	1										
9. Works well under pressure	0.07	0.01	0.05	0.02	0.18*	−0.06	0.02	0.15*	1									
10. Enthusiastic	−0.18*	0.01	0.06	0.04	0.14*	−0.21*	0.06	−0.00	0.07	1								
11. Is a critical thinker	0.02	0.03	−0.11*	0.02	0.09	−0.01	0.02	−0.05	0.09	0.06	1							
12. Is dependable	−0.07	−0.01	−0.12*	0.00	0.04	−0.02	0.08	−0.01	0.07	0.10*	−0.02	1						
13. Is anxious	−0.17*	0.10*	0.07	0.11*	−0.12*	0.05	0.08	0.07	−0.10*	−0.01	−0.14*	0.02	1					
14. Open to new experiences	−0.02	−0.10*	−0.01	−0.14*	0.21*	−0.18*	−0.15*	−0.06	0.04	0.13*	0.11*	0.03	−0.11*	1				
15. Is reserved/quiet	0.12*	0.04	−0.08	0.01	−0.29*	0.31*	0.07	−0.02	−0.02	−0.28*	0.03	−0.03	0.14*	−0.13*	1			
16. Is sympathetic/calm	−0.15*	0.08	0.02	0.13*	−0.05	−0.10*	−0.00	−0.01	−0.02	0.27*	0.04	0.05	0.16*	0.10*	0.01	1		
17. Is disorganized	0.06	0.08	0.03	0.12*	0.03	0.03	0.04	0.39*	0.00	0.01	−0.05	0.06	0.11*	−0.01	−0.04	0.01	1	
18. Is calm	0.19*	−0.03	−0.03	−0.05	0.03	0.02	−0.07	−0.08	0.05	−0.01	0.08	−0.03	−0.38*	0.11*	0.18*	0.08	−0.16*	1

*p-value < 0.05.

normality, a non-linear relationship or when ordinal variables are being used" (*11*). The highest correlation is between having a father in engineering and having any immediate family member in engineering when starting one's CEE undergraduate degree. This means that most of the time, when there was an engineer in the family, it was the respondent's father. Although most of the remaining variables are not as highly correlated, those that are follow logical trends. For example, someone who is reserved and quiet is more likely to want to work alone, whereas those who claim to be enthusiastic are more likely to prefer to work in a group setting. Likewise, respondents who agreed to being calm were less likely to agree to being anxious. In addition, those who work well under pressure are more likely to agree to procrastination tendencies than those who do not work well under pressure.

The survey responses were then tested for significance, and odds ratios were defined between different cross-cuts of the data, comparing two variables at a time. For each comparison, the first variable was something that a current CEE student would know about themselves (e.g., gender, if their father or mother is an engineer, personality trait). The second variable would be something they would like to find out (e.g., technical area of interest, role of interest). Tests for significance were run multiple times to check the relationship between two variables. For example, Pearson chi-squared tests were used to indicate whether the current technical area of alumni was significantly different by gender. The first test compared structures alumni with nonstructures alumni by gender. The second test compared transportation alumni with nontransportation alumni by gender. The other technical areas were compared with gender using an identical set up. This goodness of fit test "measures how well the observed distribution of data fits with the distribution that is expected if the variables are independent" (*12*). Groups that significantly reject the independence assumption (i.e., there is no relationship between the two variables) with 95% confidence are further analyzed using odds ratios.

It is important to note that these comparison tests require at least five samples in each cell of the chi-square matrix (*12*), which some of the smaller groups did not meet at all times. The technical field of utility engineering and surveying only had four respondents, so all of its subgroups with gender had less than five samples. Therefore, technical areas and roles with less than 5% of the samples were not compared for significance, as these groups consistently missed this minimum requirement for many of the tests. However, these observations were retained in the other analyses. That is, there is no comparison of utility engineering and surveying with nonutility engineering and surveying, but the utility engineering and surveying observations are included in the other tests (e.g., transportation with nontransportation).

Whereas the Pearson chi-squared test was used for many of the cross-cuts, it was not appropriate for testing significance in the responses to the Likert-scale personality questions. For those 14 questions, the Mann–Whitney (MW) nonparametric test (also known as the Wilcoxon rank-sum test), was used instead, as it can account for ordinal responses to surveys when respondents are asked if they agree, are neutral, or disagree with the statement (*13*). The test used for each comparison is specified in the results section.

Technical Area

Table 3 shows how alumni survey responses correlated with their technical area of interest. The *p*-values from the statistical tests are shown where significant values at the 95% level are highlighted. The table footnotes contain the odds ratio statements attributed to these tests returning a significant *p*-value of .05 or lower. For example, when the Pearson chi-squared test was performed on transportation versus non-transportation alumni responses by gender, the test returned a .02 *p*-value. Using odds ratios on this significant relationship, it was found that women in the study's sample were 1.4 times or 40% more likely to go into transportation than men.

These findings suggest that gender and personality traits may be better predictors of the technical field a person ultimately works in than their background (such as their father's or mother's profession at the beginning of the undergraduate CEE degree). Even having an engineer in the immediate family at the beginning of one's CEE degree proved to be insignificant. The significance of personality trait impact on technical area is consistent with literature. Logue et al. (*14*) conducted a study on 164 business majors and found that "49% of the variance in major satisfaction could be accounted for by a combination of vocational interest themes and personality traits."

The most distinct finding is that construction alumni have a strong preference of not wanting a desk job, whereas transportation is the most likely to want a desk job. The only other personality trait found to be significant was that transportation alumni tended to be more reserved/quiet whereas structures alumni were less reserved/quiet. The number of tests returning insignificant show how civil engineers share many of the same traits even though the technical areas are very diverse. However, the results highlight that transportation alumni may be the most unique, as three of the five significant results were with regard to comparing transportation versus nontransportation alumni.

It is also interesting that these findings overcome many limitations that university records would be unable to show when comparing technical area with gender. First, many universities do not have undergraduate students officially designate their technical track and only know that each student completed their degree program in CEE. Technical tracks are only recorded for graduate students. Second, this survey could find the ultimate placement of alumni, which, debatably, is more valuable over the technical focus while pursuing their degree. A student can specialize in one technical area, but then work for several decades in a different technical

Table 3. Chi-Squared *p*-Values and Odds Ratio Statements for Technical Area

	Statistical test	Construction	Land development	Multiple	Structures	Transportation
Gender	Pearson	0.13	0.85	0.38	0.91	0.02[a]
Father engineer	Pearson	0.95	0.74	0.72	0.92	0.17
Mother engineer	Pearson	1.00	0.70	0.18	0.87	0.10
Engineer in family	Pearson	0.51	0.49	0.44	0.68	0.43
Likes to take risks	MW	0.06	0.73	0.58	0.17	0.76
Likes to work alone	MW	0.66	0.18	0.57	0.13	0.24
Likes desk job	MW	0.00[b]	0.53	0.14	0.28	0.04[c]
Tends to procrastinate	MW	0.34	0.23	0.19	0.59	0.19
Works well under pressure	MW	0.32	0.94	0.39	0.72	0.97
Enthusiastic	MW	0.30	0.60	0.21	0.91	0.95
Is a critical thinker	MW	0.48	0.31	0.68	0.92	0.71
Is dependable	MW	0.44	0.27	0.55	0.41	0.21
Is anxious	MW	0.27	0.15	0.37	0.55	0.11
Open to new experiences	MW	0.28	0.33	0.65	0.77	0.23
Is reserved/quiet	MW	0.23	0.48	0.98	0.01[d]	0.05[e]
Is sympathetic/calm	MW	0.32	0.86	0.39	0.60	0.11
Is disorganized	MW	0.09	0.80	0.50	0.28	0.91
Is calm	MW	0.14	0.83	0.34	0.98	0.58

[a]Females are 1.4 times more likely to be in transportation than males.
[b]Nonconstruction alumni are 1.8 times more likely to like a desk job than those in construction.
[c]Transportation alumni are 1.3 times more likely to like a desk job than those not in transportation.
[d]Structures alumni are 1.6 times more likely to be reserved/quiet than those not in structures.
[e]Transportation alumni are 1.2 times more likely to be reserved/quiet than those not in transportation.

area. This is common, as several respondents working in land development commented in the survey that they did not know that technical area in CEE existed until post-graduation. In addition, as evident by the "other" category, not everyone uses their college degree in the traditional way, such as applying the analytical skills learned in CEE to the field of finance.

Respondents were then asked to select up to two characteristics that were important in their career and gave them motivation to pursue it. If a respondent chose two, each was weighted as half a vote within their technical area compared with if they had chosen one, it would be weighted as a full vote. It seems that, in general, CEE majors found the ability to grow and gain new knowledge, challenging, making a difference, and high salary/financial gain to be important aspects of their careers, potentially indicating that CEE majors have a competitive nature. However, the ability to grow and gain new knowledge was more important to those in structures and transportation, whereas high salary and financial gain were more important to construction and land development. The dynamics of each technical area's work schedule seemed to be of less importance, such as flexible working hours and whether business travel was required.

These results support current literature, as well as reiterate the importance of basing decision support on alumni careers and not on student enrollment. A previous study on freshman engineering attrition found that students "having a strong belief that engineers are paid well and that having an engineering degree helps ensure career security had a positive influence on the outcome variable, possibly implying that students who leave in poor standing may be studying engineering primarily for its financial rewards" (*15*). Had students been surveyed, it is hypothesized that job security and salary would have had increased importance when comparing the responses of the alumni.

Role

Similar to technical area, alumni career roles were not well predicted by immediate family careers at the beginning of their CEE degree, as shown in Table 4. However, gender and personality traits were significant. In addition, a high proportion of the significant comparisons occurred with respect to those in the practice/design role, indicating that the traits of alumni in that role are especially unique. When comparing Tables 3 and 4, more qualities were found to be significantly different between the roles than between the technical areas. This suggests that using questionnaires to support students in their decisions may provide more guidance when recommending roles, rather than technical area. Therefore, it would be beneficial for future research to obtain larger samples of the excluded roles (analytics, research, teaching, other) so that students can gain insight into the common characteristics in those groups.

Table 5 presents the career characteristics important to each of the different job roles. All three roles reported the

Table 4. Chi-Squared *p*-Values and Odds Ratio Statements for Role

	Statistical test	Managing	Multiple	Practice/design
Gender	Pearson	0.00[a]	0.53	0.01[b]
Father engineer	Pearson	0.64	0.60	0.44
Mother engineer	Pearson	0.11	0.80	0.15
Engineer in family	Pearson	0.51	0.11	0.45
Likes to take risks	MW	0.20	0.01[c]	0.00[d]
Likes to work alone	MW	0.56	0.32	0.09
Likes desk job	MW	0.03[e]	0.87	0.08
Tends to procrastinate	MW	0.51	0.04[f]	0.08
Works well under pressure	MW	0.36	0.40	0.13
Enthusiastic	MW	0.17	0.01[g]	0.03[h]
Is a critical thinker	MW	0.31	0.63	0.10
Is dependable	MW	0.24	0.68	0.75
Is anxious	MW	0.08	0.77	0.01[i]
Open to new experiences	MW	0.18	0.46	0.84
Is reserved/quiet	MW	0.39	0.07	0.03[j]
Is sympathetic/calm	MW	0.24	0.64	0.65
Is disorganized	MW	0.35	0.98	0.65
Is calm	MW	0.64	0.86	0.33

[a]Males are 2.7 times more likely to be in management than females.
[b]Females are 1.3 times more likely to be in practice/design than males.
[c]Multiple roles alumni are 1.3 times more likely to like to take risks than those not in multiple roles.
[d]Nonpractice/design alumni are 1.3 times more likely to like to take risks than those in practice/design.
[e]Nonmanaging alumni are 1.7 times more likely to like a desk job than those in managing.
[f]Multiple roles alumni are 1.4 times more likely to tend to procrastinate than those not in multiple roles.
[g]Multiple roles alumni are 1.1 times more likely to be enthusiastic than those not in multiple roles.
[h]Nonpractice/design alumni are 1.1 times more likely to be enthusiastic than those in practice/design.
[i]Practice/design alumni are 1.4 times more likely to be anxious than those not in practice/design.
[j]Practice/design alumni are 1.4 times more likely to be reserved/quiet than those not in practice/design.

Table 5. Important Characteristics of Career

	Managing	Multiple	Practice/design
Ability to grow and gain knowledge	17.7%	25.8%	31.7%
Ability to travel for business	0.8%	2.5%	1.1%
Challenging	21.0%	20.3%	14.2%
Flexible working hours	3.2%	5.5%	7.1%
Importance/authority/recognition	6.5%	4.7%	3.4%
Making a difference	25.0%	22.5%	16.5%
No business travel	0.0%	0.0%	0.2%
Predictable working hours	0.0%	0.8%	1.8%
High salary/financial gain	15.3%	8.9%	13.1%
Stability/job security	10.5%	8.9%	10.8%
Total	100.0%	100.0%	100.0%

Note: All values greater than or equal to 15% are highlighted.

ability to grow and gain knowledge as well as making a difference important in their careers. Those in managing had an increased rate of responding that high salary/financial gain was an important career aspect. This is correlated with findings from Table 6, where most alumni in construction are in management. Therefore, salary is important to both construction and management. Given these associations, it is difficult to create well-defined cause-and-effect relationships because it is unknown what characteristics were important in picking one's technical area and what was important in one's role. For example, an alumnus might pick construction to gain knowledge, and because it is challenging, then go into management for the increased salary. Owing to this multi-stage decision-making, it is recommended that future research includes alumni interviews to gain a more in-depth view of the decision process.

Degree Level

The last predictor analyzed was whether an alumnus earned a graduate degree in CEE (Table 7). This includes an M.S. or Ph.D. in CEE as well as a M.E. (Master's in Engineering) with a CEE focus. For this comparison, only samples with participants that started their CEE undergraduate degree between 1995 and 2010 were included, a total of 291 samples. Had all 567 samples been included, the average age difference between those who did and did not earn a graduate degree was 8 years. The significant differences were then consistent with what was expected when comparing a

Table 6. Important Characteristics of Career

	Construction	Land development	Multiple	Structures	Transportation
Ability to grow and gain knowledge	20.0%	28.3%	26.5%	35.3%	30.0%
Ability to travel for business	2.0%	0.9%	3.0%	0.0%	1.4%
Challenging	22.0%	19.1%	13.9%	22.1%	14.5%
Flexible working hours	2.0%	5.7%	6.6%	8.8%	5.5%
Importance/authority/recognition	8.0%	3.0%	4.2%	2.9%	4.1%
Making a difference	20.0%	11.7%	22.3%	20.6%	22.3%
No business travel	2.0%	0.0%	0.6%	0.0%	0.5%
Predictable working hours	0.0%	2.6%	0.6%	1.5%	0.5%
High salary/financial gain	16.0%	17.8%	8.4%	4.4%	10.0%
Stability/job security	8.0%	10.9%	13.9%	4.4%	11.4%
Total	100.0%	100.0%	100.0%	100.0%	100.0%

Note: All values greater than or equal to 15% are highlighted.

Table 7. Chi-Squared *p*-Values and Odds Ratio Statements for Graduate School.

	Statistical test	Received CEE graduate degree
Gender	Pearson	0.85
Father engineer	Pearson	0.89
Mother engineer	Pearson	0.96
Engineer in family	Pearson	0.45
Likes to take risks	MW	0.34
Likes to work alone	MW	0.07
Likes desk job	MW	0.33
Tends to procrastinate	MW	0.02[a]
Works well under pressure	MW	0.65
Enthusiastic	MW	0.99
Is a critical thinker	MW	0.85
Is dependable	MW	0.03[b]
Is anxious	MW	0.89
Open to new experiences	MW	0.07
Is reserved/quiet	MW	0.43
Is sympathetic/calm	MW	0.52
Is disorganized	MW	0.67
Is calm	MW	0.97

[a]Alumni who do not have a CEE graduate degree are 1.5 times more likely to procrastinate than those who do have one.
[b]Does not meet the five observations minimum requirement for the statistical test.

younger group with an older group and not necessarily due to differences in earning a graduate degree in CEE. After controlling for age by comparing samples in the 15-year range, the only distinguishable personality trait was that those who earn a graduate degree in CEE are significantly less likely to procrastinate. The Likert-scale responses to being dependable between the two groups returned a significant p-value, but the lack of diversity in answers resulted in too few observations in each category of the comparison.

Conclusions And Future Research

This study provides supportive evidence that personality and gender can have an impact on career decisions, including technical area and role. The family history of having an engineer in the immediate family at the beginning of one's studies was not found to be a significant predictor in alumni careers. Although there were significant differences in the types of alumni in each technical area and role, overall the career goals were mostly consistent across the different group segments, suggesting that career goals may help students pick CEE as their major, but not decisions later on in their education. Another possibility is that, consistent with previous literature, students pursuing a CEE degree with salary and job stability in mind are less likely to complete the degree program requirements. In addition, the data captured from the survey was not strong at predicting whether someone would earn a graduate degree in CEE. Although this study gives guidance to students in terms of which technical area they should consider pursuing, it seems that career role is most easily predicted as the characteristics of alumni in each role are more distinguishable from one another.

Given where significance was found, it is recommended that future studies focus on personality types (e.g., is enthusiastic) and job traits (e.g., desk job) by testing more than 14 of these preferences. Future studies may benefit from already developed tests, such as the Myers–Briggs Type Indicator to test for personality. In addition, it is recommended to focus on traits that do not have a negative connotation. For example, very few alumni disagreed or were neutral to being a critical thinker, probably because not being a critical thinker is viewed negatively. Conversely, there was great diversity in responses to whether a desk job is preferred as there was not an obvious right or wrong answer. In addition, it is recommended for future studies to implement methodologies that account for possible correlations between variables, such as regression or interview-based methodologies. Otherwise, any technical area with a high portion of managers may be incorrectly concluded to like a high salary as well. Finally, although the survey was distributed to many alumni outside of Virginia Tech, about half the responses were from Virginia Tech alumni. Future research would benefit from teaming up with other academic institutions or professional organizations to obtain a more representative sample of graduates currently in industry. This would highlight any university- or region-related impacts on the correlations as well as the robustness of the overall study conclusions.

Acknowledgments

The authors thank Angela Mills (Director of Alumni Relations at Virginia Tech) and Courtney Sakry (CEE Coordinator of Alumni and External Relations at Virginia Tech) for distributing this study's online link via email and social media. The authors also thank the 567 CEE alumni that took the time to respond to this survey.

References

1. The Whitehouse. STEM for All. 2016. https://obamawhitehouse.archives.gov/blog/2016/02/11/stem-all. Accessed July 15, 2017.
2. U.S. Bureau of Labor Statistics. *STEM Occupations: Past, Present, Future*. 2017. https://www.bls.gov/spotlight/2017/science-technology-engineering-and-mathematics-stem-occupations-past-present-and-future/pdf/science-technology-engineering-and-mathematics-stem-occupations-past-present-and-future.pdf. Accessed July 15, 2017.
3. Swartz, J. U.S. Needs More STEM Jobs to Compete: Study. *USA Today*. June 2017. https://www.usatoday.com/story/tech/talkingtech/2017/06/30/u-s-needs-more-stem-jobs-compete-study/442801001. Accessed July 15, 2017.
4. Hayden, K., Y. Ouyang, L. Scinski, B. Olszewski, and T. Bielefeldt. Increasing Student Interest and Attitudes in STEM: Professional Development and Activities to Engage and Inspire Learners. *Contemporary Issues in Technology and Teacher Education*, Vol. 11, No. 1, 2011, pp. 47-69.
5. Mendez, G., T. D. Buskirk, S. Lohr, and S. Haag. Factors Associated with Persistence in Science and Engineering Majors: An Exploratory Study using Classification Trees and Random Forests. *Journal of Engineering Education*, Vol. 97, No. 1, 2008, pp. 57-70.
6. Marra, R. M., K. A. Rodgers, D. Shen, and B. Bogue. Leaving Engineering: A Multi-Year Single Institution Study. *Journal of Engineering Education*, Vol. 101, No. 1, 2012, pp. 6-27.
7. Cantrell, P., and J. Ewing-Taylor. Exploring STEM Career Options Through Collaborative High School Seminars. *Journal of Engineering Education*, Vol. 98, No. 3, 2009, pp. 295-303.
8. Drosdeck, J. M., S. N. Osayi, L. A. Peterson, L. Yu, E. C. Ellison, and P. Muscarella. Surgeon and Nonsurgeon Personalities at Different Career Points. *Journal of Surgical Research*, Vol. 196, No. 1, 2015, pp. 60-66.
9. DataUSA. Civil Engineering: STEM Major. 2015. https://datausa.io/profile/cip/1408/#demographics. Accessed July 15, 2017.
10. American Society of Civil Engineers. Technical Areas. 2017. http://www.asce.org/routing-page/technical-areas. Accessed July 15, 2017.
11. Laerd Statistics. Spearman's Correlation using Stata. 2013. https://statistics.laerd.com/stata-tutorials/spearmans-correlation-using-stata.php. Accessed July 15, 2017.
12. University of Pennsylvania. Tutorial: Pearson's Chi-square Test for Independence. 2008. http://www.ling.upenn.edu/~clight/chisquared.htm. Accessed July 15, 2017.
13. Laerd Statistics. Mann–Whitney *U* Test using SPSS Statistics. 2013. https://statistics.laerd.com/spss-tutorials/mann-whitney-u-test-using-spss-statistics.php. Accessed July 15, 2017.
14. Logue, C. T., J. W. Lounsbury, A. Gupta, and F. T. Leong. Vocational Interest Themes and Personality Traits in Relation to College Major Satisfaction of Business Students. *Journal of Career Development*, Vol. 33, No. 3, 2007, pp. 269-295.
15. Besterfield-Sacre, M., C. J. Atman, and L. J. Shuman. Characteristics of Freshman Engineering Students: Models for Determining Student Attrition in Engineering. *Journal of Engineering Education*, Vol. 86, No. 2, 1997, pp. 139-149.

The Standing Committee on Transportation Education and Training (ABG20) peer-reviewed this paper (18-03163).

Article

TRR
JOURNAL OF THE TRANSPORTATION RESEARCH BOARD

Transportation Research Record
2018, Vol. 2672(13) 10–18

DOI: 10.1177/0361198118758057
journals.sagepub.com/home/trr
SAGE

Evaluation of the Transportation Research Board Minority Student Fellows Program

Richard Willson[1] and Karen Febey[2]

Abstract

This paper evaluates the Transportation Research Board (TRB) Minority Student Fellows Program (Program). Initiated in 2010, the Program supports students from select Historically Black Colleges and Universities, Hispanic-Serving Institutions, and Native American Pacific Islander-Serving Institutions to attend TRB's Annual Meeting and present a student-authored research paper. The Program is intended to increase involvement of those groups in TRB and the transportation profession. This evaluation considers the Program's performance against stated and implicit goals using a survey of mentees; qualitative reflections by the Program manager, a mentor, and a mentee; and indirect outcome metrics. Conclusions are that the Program was successful in increasing exposure to TRB, contributing to the growth of fellows as transportation professionals, and encouraging ongoing TRB participation. With regard to operations, the Program achieved visibility and attracted external funding, which bodes well for its sustainability. Lessons learned include crafting appropriate rules for participation and finding the right balance between critique and support in the paper review process. There are opportunities to extend the mentoring experience with continued involvement of past fellows.

Workforce diversity benefits the transportation profession by bringing a broader spectrum of viewpoints to the workplace, providing insights into the transportation needs of underrepresented communities, and ensuring that the transportation professions mirror the characteristics of the population. While diversity encompasses many characteristics of people's identity—such as race, ethnicity, gender, sexual orientation, socio-economic status, physical ability, and religious and political beliefs, the Transportation Research Board (TRB) chose to focus its program on racial and ethnic minority populations that are underrepresented in the transportation field. These populations include those who identify as African-American, Hispanic, American Indian/Alaska Native, and Pacific Islanders—who are underrepresented minorities (URM) in the transportation field.

These underrepresented populations account for just over 13% of civil engineers in the US workforce (*1*), yet comprise about 30% of the US population (*2*). Among all undergraduate science, technology, engineering, and mathematics (STEM) degree recipients, those who are URM students mirrors the level of involvement of these same groups on TRB panels, committees, and as attendees at the TRB Annual Meeting (*3*).

This paper evaluates the TRB Minority Student Fellows Program (Program). The Program provides funding and other support for students from select Historically Black Colleges and Universities, Hispanic-Serving Institutions, and a Native American Pacific Islander-Serving Institution to attend TRB's Annual Meeting and present a student-authored research paper at a poster or lectern session. The Annual Meeting, held every January in Washington, DC, is the largest transportation research conference in the world, hosting 13,000+ attendees who attend more than 5,000 presentations in over 800 sessions and workshops.

About the Program

The origins of the Program date back to TRB's 2007 Strategic Plan, in which the TRB Executive Committee outlined efforts to increase diversity through more involvement of women, minorities, and young professionals in the Annual Meeting, TRB, and the transportation profession (*4*). One of the ways through which the Committee sought to increase diversity was through engagement with Hispanic-Serving

[1]Department of Urban and Regional Planning, California State Polytechnic University, Pomona, CA
[2]Transportation Research Board, The National Academies of Science, Engineering, and Medicine, Washington, D.C.

Corresponding Author:
Address correspondence to Richard Willson: rwwillson@cpp.edu

Institutions (HSIs) and Historically Black Colleges and Universities (HBCUs). In developing the structure of the Program, TRB staff and Executive Committee members outlined a program to bring students from HSIs and HBCUs to the Annual Meeting and that would include the following components: (1) a preference to support faculty members from participating institutions to oversee fellows' research and writing prior to the Meeting and provide them mentoring while at the Meeting; (2) a need for local sponsors to support a long-term funding structure; (3) participating school ownership of the selection process; and (4) involvement of fellows in the TRB paper review process, albeit with guaranteed acceptance conditional on paper revisions (*5*).

In addition to funding fellows' Annual Meeting expenses, the Program provides partial funding support for their faculty mentors so that the mentors can help fellows navigate the Meeting, attend events together, and provide support for fellows during their research presentations.

While formal program activities take place at the Annual Meeting, preparation before the Meeting and TRB involvement after the Meeting also offer students opportunities to develop career-specific skills, form relationships through networking and mentoring, and learn about the state of practice in their fields. Before the Annual Meeting, students work closely with faculty mentors at their school to carry out research and draft a research paper, respond to peer review comments, prepare oral presentations of their research, and map out their schedule for the Meeting. During the Annual Meeting, TRB arranges networking events for the fellows (including a reception and luncheon) so that they can learn about involvement in TRB and meet TRB staff and other professionals in their fields. After the Annual Meeting, fellows are strongly encouraged to involve themselves in TRB's committee and panel structure to develop their expertise, continue professional networks, and learn about developments in their fields.

Figure 1 summarizes the development and ongoing evolution of the Program, indicating periods of planning, implementation and growth, and stability and evaluation.

The Program has supported 104 students since its inception in 2010. Of those fellows, more than half identify as African-American (62%), followed by 38% as Hispanic, and 4% as Native American/American Indian. The group comprises 55% men and 45% women. Although there is no preference for women in the selection process, the Program has the added effect of increasing gender diversity in the transportation profession.

The fellows' course of study is predominantly engineering (78%), with planning representing 17%, and other fields such as supply chain management and airway science management representing 5%. Engineering includes the subfields of civil, structural, mechanical, environmental, and geotechnical. The weighting toward engineering is reflective of the overall composition in TRB, and addresses the lack of underrepresented minorities in engineering.

Eligible institutions have latitude to select their own fellows as long as the students meet basic program criteria, which include having a transportation-related major, submitting their research papers by the August 1 deadline, responding to peer review comments, and being able to participate in all Program activities. In addition, institutions have the latitude to select the faculty member who will accompany their students to the Annual Meeting and act in a mentoring role. Since participating institutions operate their own nomination and selection process, fellows come from a diversity of academic levels: 43% undergraduate, 47% Master's students, and 10% PhD students or candidates. Given this composition, the Program encompasses younger students with little exposure to the research process as well as those students with substantial prior research experience.

Literature

The benefits of mentoring on students' academic and professional success are well established. Those who are mentored often assimilate better into their workplace cultures, have a more extensive professional network, and make more substantive contributions to their organizations and their field as a whole (*6,7*). However, students and young professionals from underrepresented minority populations are often less likely to receive mentoring as they develop as STEM professionals, but when they do receive mentoring, they benefit greatly from such relationships (*7,8*).

There is no one-size-fits approach to mentoring, and there is no one type of mentor. Mentoring models include new-hire mentoring and peer-to-peer mentoring at an organization, as well as cross-generational mentoring (*9*). Mentors provide educational and career advice, offer an apprenticeship-like relationship to complete a particular task, and give general life coaching. Mentoring can range from one meeting to a relationship that spans decades, and mentors can be professors, colleagues, other experienced professionals, or those encountered through formal mentoring programs (*7,9,10*).

Mentoring can take place through one-time events like conferences which often include activities such as networking events and career panels (*7*). Less formal networking with a broad array of professionals can influence students' professional development, help them understand how practice and research inform each another, and assist them with developing research writing and oral presentation skills (*11*).

Generational differences may be relevant in developing mentoring programs. Millennials may anticipate career paths, styles of working and communication, and workplace values that are different than those of previous generations. These differences have implications for the design of mentoring programs (*10*). For example, addressing generational differences directly, as they pertain to communication and hierarchy, could benefit mentoring programs.

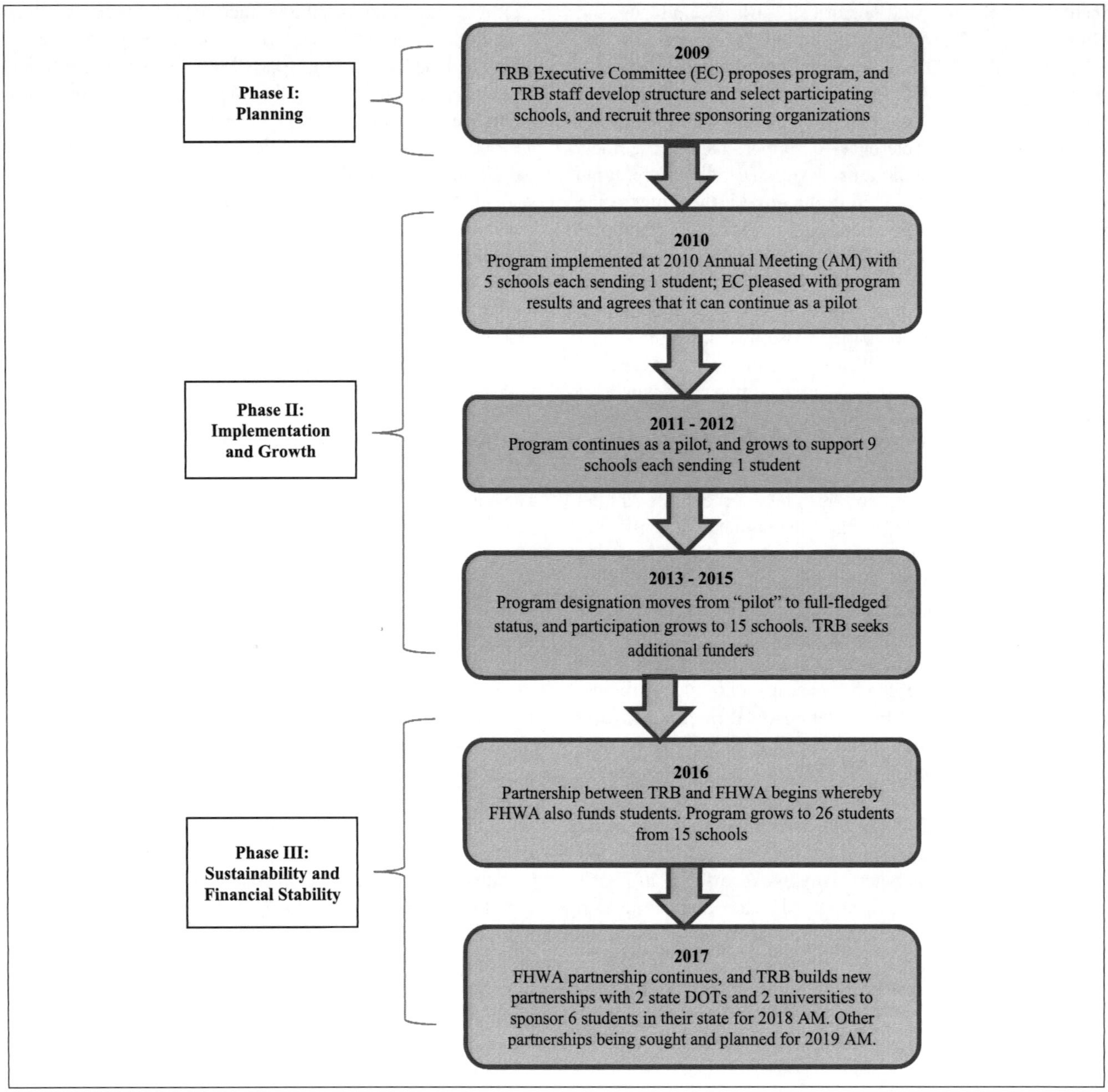

Figure 1. Development phases of TRB Minority Student Fellows Program.

Methodology

The research design is a case study that allows for an in-depth analysis bounded by both time and activity (*12,13*). To carry out this study, qualitative and quantitative data are drawn from three primary sources.

First, a survey was administered to all fellows from the 2010–2016 cohorts. TRB staff administered the survey in April 2016 as part of a formative evaluation to learn more about (1) fellows' accomplishments since Program participation; (2) perceptions of how the Program influenced their academic and career development; (3) the extent of their involvement (if any) in TRB committees and panels; and (4) suggestions for Program improvements. The responses also provide information about fellows' careers since graduating, the role that networking and mentoring played in their development, and how fellows came to decide on a career in transportation. As the second source of information, the TRB program manager, a faculty mentor from a participating institution, and a former fellow provide their reflections on the Program's operations, effects, and ideas of how the Program can be enhanced. Finally, other metrics on Program performance, such as trends in participation, external funding, and Program publicity are included in the evaluation—which are based on documentation presented during TRB's Executive Committee meetings. While results of this evaluation may not be transferable to all programs, this research can inform those designing similar programs in other fields.

Survey Analysis

The survey of Program alumni was conducted in April 2016. It received 50 responses, representing 63% of the alumni at that time. Responses from cohort years 2010 through 2016 were received, with 39% of responses from the 2015/16 year. Consequently, the responses were weighed toward recent participants who are in the early stages of their careers. The responses are generally representative of the course of study and demographic composition of all alumni of the Fellows Program.

Fellows reported on the kinds of mentoring they received from their faculty mentor in the areas of research and career development. In this closed-ended question that allowed fellows to select more than one response, 35 (70%) reported that their research-related mentoring included exposure to the research process through assistance with research techniques, and 32 (64%) reported that such mentoring was through the paper writing/peer review process. An additional open-ended question also asked about the type of mentoring that was most useful. These narratives were coded on common themes, with the most frequent research-related comments addressing how to write a technical/research paper (8 responses), guidance with research methods (4 responses), research expectations and feedback (3 responses), and preparation of materials for presentation (3 responses).

An open-ended question asked about changes that fellows suggest for the Program. Of the 36 responses received, 16 said "no changes." The remaining 20 responses included suggestions for more formalized mentoring, a fellows career expo, more orientation to the Program, and more interactions with mentors. One student felt the term "minority" should be removed from the title, citing a concern that it could alienate potential fellows.

While the initial charge for the Program outlined goals for its operation, detailed goals for fellow outcomes were not articulated in TRB's Executive Committee Meeting minutes. The following organizes the key survey results into a series of themes based on the authors' understanding of formal and implicit outcome goals that have guided the Program.

Exposure to TRB and the Transportation Industry

The Program offered most fellows a first-time TRB experience—only 9% had attended the Annual Meeting before. While not measured as part of the survey, fellow anecdotes also suggest that it was the first trip to Washington DC for many.

Fellows' Growth as Young Transportation Professionals

Fellow were asked an open-ended question about the most beneficial parts of the Program, including new strengths developed as a professional. Their answers were coded into key themes. The most commonly cited benefit was networking with other professionals and peers (14 comments). Fellows mentioned that they enjoyed meeting the wide range of transportation subject matter professionals, students from other schools, and transportation professionals from other states and countries.

The second most commonly cited benefit was experience in making presentations (13 comments). All fellows presented in a lectern or poster event. One respondent wrote "Feeling vulnerable and ready for criticism during the poster session helped me develop my presentation skills, confidence, and overall provided me with a sense of achievement after my presentation session was up."

The third most-frequent set of comments relates to conference attendance broadening fellows' perspective on the transportation field (12 comments). Fellows noted that their "eyes were opened" and that they learned of other practices in the transportation field. Related to this, comments were also made about having access and exposure to other professionals (7 comments).

The research, paper writing, and feedback process was mentioned in seven comments. Other topics receiving multiple comments include seeing others' research (4 comments), and motivation, minority empowerment, and confidence (4 comments).

Fellows' Ongoing Engagement in TRB and Lasting Effects

Many fellows returned to the TRB Annual Meeting in the following years, with 31% responding that they had attended after being fellows. Given the expense of conference travel, limited agency travel budgets for junior staff, and the fact that not all fellows ended up working in transportation, this seems an impressive level of participation. Of the 50 survey respondents, a substantial number of respondents indicated involvement beyond attendance: two are members of a Standing Committee or Subcommittee, 18 are a friend of a Standing Committee or Subcommittee, and three have been paper reviewers. The reasons that fellows offer to explain continuing involvement include networking (26 responses), developing skills and knowledge (24 responses), learning more about the field (22 responses), learning about career opportunities (19 responses), and adding professional activities to their résumés (17 responses).

Career development elements were queried in a closed-ended mentoring question that asked about types of mentoring that fellows received from their mentors while preparing for the Annual Meeting. Respondents indicated that they utilized their mentors in seeking advice to make the most of their time while at the Annual Meeting (22 responses), networking (16 responses), and career planning (11 responses). In an open-ended question about the extent to which the mentoring was helpful to fellows in their studies or careers, comments about benefits were categorized as one-on-one connections with their faculty mentor for guidance (5 responses), expectations of work and feedback (3 responses), conference and TRB Annual Meeting guidance (3 responses), career navigation (2 responses), and contacts/networking (2 responses).

Three-quarters of respondents (76%) consider themselves to be currently working in the field of transportation. Three respondents pursued a PhD following the Program, while six of the undergraduate fellows who responded pursued Master's degrees. Fellows reported employment in a wide variety of transportation sectors, with "consulting firm" and "enrolled in school" being the most common responses (30% each).

Many fellows report meeting other professionals at the conference and keeping in touch (40%). In terms of fostering community among fellows, over 35% said they kept in touch with other fellows. In terms of professional development, a significant number of respondents obtained an Engineer in Training/Engineering Intern certification.

Reflections from the Program Manager, a Mentor, and a Mentee

The mentoring provided in the Program is more comprehensive than simply providing paper or career advice. Mentees embark on a journey with their mentors—in writing their paper, presenting it, attending the conference, identifying Annual Meeting events that align with their academic and career goals, and networking. Their needs vary according to their academic and life experience. Accordingly, qualitative impressions of the experience are an important part of this evaluation, shedding light on dimensions that the survey may not reveal. The following provides reflections from the Program manager (author 2), a mentor (author 1), and a mentee.

Program Manager, Karen Febey

I have been the Program manager of the Fellows Program since 2013. In that role, I ensure the financial viability of the Program, work to help engage fellows in the transportation profession through the Annual Meeting and TRB, and handle all Program logistics. Specifically, I have developed partnerships with outside organizations to sponsor fellows, matched fellows with Annual Meeting mentors (in addition to their faculty mentors), planned networking opportunities for them at the Annual Meeting, and made them aware of how they can become involved with TRB committees and panels.

For many students, the Annual Meeting is their first contact with the transportation profession outside of their schools. My role as Program manager is not just to handle practical logistics but also to ensure that fellows prepare a quality paper and research presentation, know how best to utilize a mentor, efficiently schedule their time at the Annual Meeting, and understand opportunities to become involved in TRB committees and panels.

As the Program has grown, there has been a need to formalize rules and procedures regarding criteria for schools and fellows' participation, expectations for mentors, and Program activities for fellows. The goal is to ensure that mentors and fellows follow the same rules for Program participation, there is financial accountability and transparency, and there are consequences for not adhering to the rules. For example, fellows and mentors are provided with the dates and times of the fellows' required events well in advance and are informed that their schools' continued eligibility depends on their fellows' participation in all of those events and on presenting their research. In one case, a school lost eligibility after the fellow from that school purposely scheduled a flight to leave the day before her research presentation and a required event.

However, a consideration in formalizing rules and procedures is whether such rules and procedures could be interpreted as pejorative or overly restrictive in a way that would discourage schools from sending fellows and potential sponsors from funding fellows. From my view, I want TRB and all sponsors to know how every dollar of Program funds is spent and have assurances that their funds are being used as intended—for fellows' participation in all Program activities and the Annual Meeting.

For TRB as an organization, the Program has encouraged staff to think about ways to partner with other minority-serving institutions and organizations that represent those underrepresented in the transportation profession. And for those in the transportation community, the Program serves as evidence of TRB's efforts in direct engagement and outreach with those whose perspectives are needed to advance the transportation profession.

Mentor, Richard Willson

I have been mentoring Program participants since Program inception in 2010, including graduate and undergraduate urban and regional planning students at California State Polytechnic University, Pomona (CPP). CPP is an HSI that is noted for helping students climb the economic ladder, so the Program is a great fit with our mission. The Program's elevation of research, faculty mentoring, and funding support has made a large impact.

The Program had a varied range of impacts on students selected as fellows from CPP. The CPP fellow who has gone the farthest academically, post-Program, was a Master's student who was likely on a track to a PhD before being a fellow. After participating, however, the recognition he received and the research he completed helped him gain acceptance to a PhD program in City and Regional Planning. Another fellow was already a transportation planner, working on a Master's degree in the evening, and is now involved in a TRB Standing Committee. In these cases, the Program solidified and strengthened an already-established path. In other cases, the Program has been an eye-opening and confidence-building experience for other students we select as fellows, as the experience shows them that "they can do it" and that transportation planning offers varied opportunities. While it is appropriate to measure specific outcomes of the Program, there is also value in allowing a "big tent" view of possible impacts. In addition, our fellows may not realize how much the Program influenced them until decades later.

In my experience, the fellow's research paper often flows out of a paper already started in a class. Still, there is significant stress for the mentor—will the fellow deliver with quality and timeliness, and how will that reflect on the Program? Mentors must decide whether to play a directive role, and thereby maximize the quality of the project, or let the fellow struggle to define the problem and research question. This second path can provide a deeper learning experience, but a less-polished paper. In my experience, the right place on this continuum of approaches is different for each student.

Perhaps the most valuable period in the mentoring experience is when the fellows I am mentoring receive the peer review comments. This is when the mentor puts the process in perspective and explains practical responses. Fellows are rightly uncomfortable when the criticism is harsh, contradictory, or asks for things the analysis cannot provide. Yet this is a valuable learning experience— the author makes a good faith response and acknowledges study limitations. At the conference, most fellows are nervous about their research presentation, but after it is complete, they are exhausted and inspired. They are enthused, and sometimes amazed, by finding like-minded people who take interest in their work, respect it, and want to discuss it in a constructive manner.

Fellows also tell me that the Program makes them aware of the vast scope of the transportation field, the international dimension of the field, the Washington DC policy environment, and the structure and functioning of a large organization like TRB.

The Program's impacts on the student population at the participating institutions is greater than the benefits that flow to the individual fellow. TRB's mission is made known when the Program is announced, and in our case, numerous other students now attend the Annual Meeting. TRB was not generally on the urban and regional planning students' radar before the Program. Fellows' sharing of experiences on social media expands this awareness even further.

As a teacher and scholar, the Program has focused my attention and attendance on TRB and publishing in the *Transportation Research Record*. There is efficiency in mentoring a student on his or her paper and preparing my own TRB paper with the same the deadlines and event schedule.

For my academic department, the Program is among the top enrichment programs we offer our students. It provides multiple benefits that support our mission—diversifying the transportation profession, raising awareness of transportation planning, encouraging student research, and most important, recognizing the value of the mentor/mentee relationship.

Former Fellow, Christine Wipfli, Florida International University

I was a fellow in the 2015 cohort, participating as a student from Florida International University. As a student pursuing a bachelor's degree in engineering, the opportunity for exposure to industry professionals and development of interpersonal and communication skills was invaluable. The Program offered a unique experience for students to grow both personally and professionally, while gaining valuable skills to bolster the resume before graduating.

Throughout each stage of the Program I strengthened the skill sets which are essential for a successful career. At the onset of my project, I honed my analytical skills and time management abilities while conducting my research. The research itself was challenging, as it required great effort to efficiently review vast amounts of literature while maintaining a clear focus on how the information would support the overall structure of the report. Drafting the final report gave me an opportunity to improve my writing skills, and also a chance to practice transferring technical text and industry-specific terminology into a writing style that would also be engaging and understandable to the non-expert.

Upon completing the final report, the peer review process was an invaluable opportunity to get feedback from industry professionals on the report itself and suggestions for improvement. This feedback proved to be extremely helpful, as it allowed me to get a clear perspective on my progress and the overall quality of my project. The most impactful and rewarding part of the experience came when I had the opportunity to present my hard work and research to the conference attendees who are at the forefront of the transportation industry.

For new TRB fellows seeking advice, I would suggest investing a great deal of time and effort selecting the optimal research topic. When thinking back on my experience, I had a difficult time selecting a topic that was most relevant to the field, feasible as a short-term research project, engaging in content, and abundant in the data and information available. When I arrived at the conference I was uncertain how relevant or interesting my topic would be to the conference attendees. To new fellows, I would suggest to consider selecting two to three potential topics. Then, I would suggest reaching out to an industry expert (perhaps a transportation professor or even the TRB peer reviewers, if possible) to narrow down the options based on feedback about which topic would be most relevant for the industry and conference attendees at that point in time.

Table 1. Program Operations Findings and Suggestions

Criterion	Findings	Suggestions
Financial stability	Program has grown and attracted external funding partners. Starting with three schools, the Program now will have 16 schools and 26 fellows in 2018.	TRB can share results and success stories with sponsors to maintain interest. Target state Departments of Transportation as new sponsors.
Fellows' participation level	Occasional inconsistent participation has been improved with clear communication of expectations.	Existing communications processes are working, but can of course be tweaked every year.
Program branding and awareness	Program has visibility at participating institutions, in conference materials, and the Chairman's lunch. Articles in *TR News*, on the Program's webpage, and in university and trade magazines share the word.	Poster and lectern sessions by fellows could be highlighted in the conference program. TRB leadership could consider if there is sensitivity to the "minority" in program's title and revise if this is an issue. TRB could share results with other STEM programs.
Impact of diversity in TRB and the profession	Because the Program is small relative to TRB, it is difficult to attribute any before/after difference of diversity to the Program. Survey results and qualitative assessments indicate that fellows may not have otherwise attended. Growth in the number of participating schools, external sponsorships, and ongoing fellow involvement in committee and panels indicates a positive impact.	Repeat fellow survey in 2018. Additional assessment tools could include (1) surveying participating schools about awareness, experiences, and impacts; (2) surveying transportation employers to determine if they recognize the Program and if it has affected their efforts to hire underrepresented minorities; and (3) tracking fellows' career progressions.

I feel very fortunate for the opportunity to have been a TRB fellow. Looking back on my experience, it is easy to see the importance of the skills that I developed, and how it will be beneficial in my future profession. As I continued my career working at the International Atomic Energy Agency, those project management skills, and the analytical and presentation skills that I developed in the fellowship became even more valued. In my current position conducting research for the US Department of Energy, research and technical writing skills play a crucial role in the work that I do. I would highly recommend this TRB Minority Student Fellows Program to undergraduate and graduate students who are interested in both challenging themselves and broadening their skill sets in the STEM industry.

Evaluation

Table 1 lists four *Program operations* criteria, based on the official and implied goals of the Program. They include financial stability, participation expectations, Program branding and awareness, and impact on diversity. The findings draw on the survey, the qualitative observations, and data on Program characteristics and outreach.

Table 2 lists the four main desired *outcomes for fellows* themselves. They include exposure to TRB, ability to navigate TRB, growth as a transportation professional, and ongoing TRB participation.

Conclusions, Study Limitations, and Recommendations

The following provides conclusions, comments on assessment methods and limitations, and recommendations.

Conclusions

Tables 1 and 2 summarize the key conclusions about the Program. It has met its goals and is in a strong position to grow in the future. As noted, the benefits are not just to fellows who attend the Annual Meeting, but also flow to faculty mentors, the fellow's institution, and TRB as an organization.

Mentoring and the personal connections that fellows made were often the most significant aspects of their experience, mentioned more than the actual sessions they attended and the subject matter expertise they acquired during sessions. These direct mentoring relationships and connections formed through networking are the core benefits of the Program. The Program is worthy of replication in other technical fields.

Assessment Methods and Limitations

While evaluation of enrichment programs is essential, it is important to recognize that experimental design conditions cannot exist as there is no equivalent control group and factors explaining success are multifaceted. Evaluations should use both quantitative and qualitative methods. These could include quantitative rating scales on surveys but also qualitative storytelling methods that ferret fellows' "stories" about how their backgrounds and professional goals informed their experiences as fellows. However, even when using such mixed methods, it can be challenging to determine the impacts of the Program from fellows who have had vastly different levels of research, conference, and work experience, and who are at varying stages in their academic careers (i.e., undergraduate, Master's student, PhD student/candidate). In addition, the Program is small in comparison

Table 2. Fellow Outcomes Findings and Suggestions

Criterion	Findings	Suggestions
Exposure to TRB	Fellows report an "eye-opening" experience regarding their view of the transportation profession: meeting seasoned professionals, participating in committees, and learning about the international dimensions of the profession. The mentor reflection confirms this effect.	Encourage fellows to return to the Annual Meeting. Ensure fellows know of possible funding options so that they can return to TRB.
Ability to navigate TRB (and other professional settings)	The Program introduces the fellows to TRB as an organization and the Annual Meeting. It demystifies the research and peer review process, and empowers them to enter the transportation profession.	The Annual Meeting is the first conference experience of many fellows, so guidance is needed. TRB could create a manual for fellows on meeting TRB standards, responding to peer reviewer comments, and maximizing time at the Annual Meeting. While this information is currently provided by mentors, a manual could serve as a back-up to the mentor's advice. A manual for first-time mentors could assist them provide the best advice.
Growth as transportation professionals	Paper preparation and presentation process is favorably perceived. Other reported benefits include networking, mentoring, presentation, and exposure to the field.	To set fellows up for successful mentor relationships, provide them with detailed guidance on how to maximize the relationship—even if it just involves a couple meetings.
Ongoing TRB participation	Significant numbers of fellows are returning to TRB, both as students and as young professionals. Some are involved in committees and panels.	Expand the number of fellows who become involved in TRB's standing committees and panels. Engage past fellows to act as professional mentors to each incoming group of fellows.

to other STEM educational programs so it may not be possible to infer that any impacts from TRB's Fellows Program would be replicated in other similar programs. Finally, the paper authors are involved in the Program as administrator and as a faculty mentor. Therefore, bias could be a limitation as the authors may not have been able to provide as critical an analysis of the Program as someone who was an external evaluator.

Recommendations

As the Program grows, and as TRB seeks more sponsors/funders, it is important to have a balance of rules/guidelines that ensure fairness for all and provide financial transparency, with flexibility that allows the Program to adapt and improve.

To supplement the mentoring that students receive from their faculty mentor, more attention could be paid to engage more fellows in the mentoring initiative that pairs fellows with established professionals in their desired field while at the Annual Meeting (currently optional for fellows, and only a handful of fellows request such a mentor each year). In addition, an educational document could be created to give students guidance on how to seek out mentors, the importance of mentors, and information on different types of mentoring relationships. This could support fellows requesting an Annual Meeting mentor and provide advice to use in their career as they seek mentors on their own.

The Program benefits those who have already selected transportation as a professional interest and those who are high achievers. There is an opportunity to develop additional programs that seek to engage high school and college students in the transportation field, to bring a more diverse cohort *into* transportation programs.

References

1. National Science Foundation. *U.S. Scientists and Engineers, By All Degree Levels, Level of Highest Degree, Ethnicity, and Race: 2013*. National Center for Science and Engineering Statistics, Arlington, Va., 2013. https://ncsesdata.nsf.gov/us-workforce/2013/html/SES2013_DST_06_0.html. Accessed 19 June, 2017.
2. U.S. Census Bureau. *Profile of General Population and Housing Characteristics: 2010*. Government Printing Office, Washington, D.C., 2010. https://factfinder.census.gov/faces/tableservices/jsf/pages/productview.xhtml?pid=DEC_10_DP_DPDP1&src=pt. Accessed 19 June, 2016.
3. Transportation Research Board. Participation of Committee and Panel Volunteers by Race, Ethnicity, 2017. Unpublished raw data.
4. Transportation Research Board. TRB Executive Committee, Strategic Plan, 2007. Unpublished raw data.
5. Transportation Research Board. TRB Executive Committee Meeting Minutes, 2008, 2009. Unpublished raw data.
6. Johnson, W. B. *On Being a Mentor: A Guide for Higher Education Faculty*, 2nd ed. Routledge, New York, 2016.
7. Packard, B. W. *Successful STEM Mentoring Initiatives for Underrepresented Students: A Research-Based Guide for Faculty and Administrators*. Stylus, Sterling, Va., 2016.
8. O'Neil, K. *Mentoring Is the Key to Increasing Minority and Women's Participation in STEM Education, Researchers Say at the Emerging Researchers Network Conference*. AAAS,

2016. https://www.aaas.org/news/mentoring-key-increasing-minority-and-women-s-participation-stem-education-researchers-say. Accessed 19 June, 2017. Washington. DC.
9. Blanchard, K., and C. Diaz-Ortiz. *One Minute Mentoring: How to Find and Work with a Mentor—and Why You'll Benefit from Being One*. William Morrow, New York, 2018.
10. Willson, R. *A Guide for the Idealist: Launching and Navigating Your Planning Career*. Routledge, New York, 2017.
11. Mata, H., T. P. Latham, and Y. Ransome. Benefits of Professional Organization Membership and Participation in National Conferences: Considerations for Students and New Professionals. *Health Promotion Practice*, Vol. 11, No. 4, 2010, pp. 450–453.
12. Merriam, S. B. *Case Study Research in Education: A Qualitative Approach*. Jossey-Bass, San Francisco, 1988.
13. Yin, R. K. *Case Study Research: Design and Methods*. Sage, Newbury Park, Calif., 1989.

The Standing Committee on Transportation Education and Training (ABG20) peer-reviewed this paper (18-00964).

Article

TRR
JOURNAL OF THE TRANSPORTATION RESEARCH BOARD

Transportation Research Record
2018, Vol. 2672(13) 19–28

DOI: 10.1177/0361198118781394
journals.sagepub.com/home/trr
SAGE

Leveraging State DOT Technology and Practice for Cost-Effective Pavement Management by Local Government Authorities

Zhaohua Wang[1], Yichang (James) Tsai[2,3], and Binh Bui[4]

Abstract

To improve management of the pavement assets in Georgia's local government authorities (LGs), there is a need to leverage the existing pavement management and maintenance technology and practice in the Georgia Department of Transportation (GDOT). However, due to the difference in organizational structure and resource limitation of LGs, GDOT's technology cannot be directly adopted. To address this issue, a research team at Georgia Institute of Technology (Georgia Tech) developed the Computerized Pavement Condition Evaluation System for Counties and Cities (COPACES-CC) that is based on GDOT's current practice on pavement management and maintenance. To facilitate the technology transfer (T^2) from GDOT to Georgia's LGs, the Georgia Tech research team has collaborated with GDOT's Office of Research, Local Technical Assistance Program, Office of Local Grants, and Office of Maintenance. LGs' decision makers and stakeholders were also engaged in this T^2 activity. This paper presents the research need and development of COPACES-CC, a guided T^2 process that was adopted in three-stage development and implementation strategies, and the key factors that are related to the components of a guided T^2. Conclusions and future recommendations are offered. It is hoped that the successful experience in the T^2 of COPACES-CC could be useful and beneficial to other state DOTs and LGs.

Local government authorities (LGs) manage three-quarters of the total 4 billion miles of roadway and more than half of the nearly 600,000 bridges in the U.S.A. (*1*). These are critical transportation infrastructure assets to support the mobility, economy, and homeland security of the local communities and the nation. To maintain the aging transportation infrastructure in good repair under shrinking budgets, state Departments of Transportation (DOTs) have adopted asset management systems to conduct cost-effective maintenance, rehabilitation, and reconstruction (MR&R). However, most LGs still rely on the knowledge and experience of individual staff to manage and make decisions on transportation infrastructure MR&R because they lack the resources and technical expertise to adopt a more systematic approach (*2*, *3*). This makes it difficult for LGs to maximize the benefit of their stringent transportation budgets or to demonstrate accountability to legislators, the public, and other stakeholders.

Though various commercial software packages for managing road pavements are available, the lack of support from a DOT in the same state hinders their widespread adoption by LGs, since they normally require heavy dependence on vendors and/or consultants in the private sector. The existing commercial software packages were designed for a broad market. Thus, they often use some standard pavement distress protocols, such as the one defined in ASTM D6433 (*4*) or its derivatives. However, there is no uniform distress protocol adopted by all the state DOTs in the U.S.A. In contrast, almost every state DOT has developed its own pavement distress protocol and has utilized the corresponding pavement condition data to manage and maintain its pavements statewide. Because the knowledge of pavement management and MR&R decision making has been accumulated through a long history of utilizing a self-developed pavement distress protocol in a highway agency, it is costly to shift to a new distress protocol in considering the extra budget need and the availability of local expertise. Normally, if an LG plans to use commercial software for managing its pavements, it needs to hire a consultant company to set up the system, collect distress data, and recommend network-level and project-level project selection, treatment selection, and budget allocation. In the meantime, the LG also needs to train its pavement engineers to become familiar with the new pavement distress protocol and its linkage to pavement MR&R. Because

[1]Center for Spatial Planning Analytics and Visualization, Georgia Institute of Technology, Atlanta, GA
[2]School of Civil and Environmental Engineering, Georgia Institute of Technology, Atlanta, GA
[3]Chang'an University, Xi'an, Shaanxi, P.R. China
[4]Office of Research, Georgia Department of Transportation, Forest Park, GA

Corresponding Author:
Address correspondence to Zhaohua Wang: zhaohua.wang@coa.gatech.edu

state DOTs, for example, the Georgia DOT (GDOT), use their own pavement distress protocols, it is difficult for an LG to leverage the resources within its local state DOT, such as support from the Local Technical Assistance Program (LTAP) and the state DOT's pavement management and maintenance technology and practice. In addition, many retired pavement engineers are hired/recruited by LGs or consultant companies. Their knowledge and experience is of great value to promote the LG's pavement management and maintenance. However, if a new distress protocol is adopted, these resources can no longer be utilized as effectively.

To address the above issue, it would be wise to transfer a state DOT's technology on pavement management and maintenance, that is, software tools, practice, and knowledge, to the LGs in the same state. However, it is not a trivial task. Challenges exist and need to be addressed for a successful implementation. Using Georgia as an example, there are 159 counties and more than 500 cities within the state. It is difficult to develop a pavement management system (PMS) suitable for all the LGs. It is also infeasible to develop a customized tool for each of them. In addition, the linear referencing systems (LRS) used by GDOT for defining project locations are not currently adopted by all the LGs, if any. Thus, a flexible means of location reference is needed. Since Federal Highway Administration (FHWA) has required the state DOTs to submit their LRS to include all public roads in the Highway Performance Management System (HPMS) (*5*), it is hoped that the LRS for GDOT and all the LGs in Georgia can be aligned in the future.

As a case study of technology transfer (T^2), this paper presents recent progress in transferring GDOT's pavement management technology and practice to Georgia's LGs by addressing the above issues. It is hoped that the experience gained and lessons learned will be valuable for other state DOTs and LTAPs that are seeking to transfer the readily available transportation innovations and tools to the LGs.

Research Need and Development to Transfer GDOT's Pavement Management Technology and Practice to Local Governments

Needs of Pavement Management in a Local Government Authority

GDOT has been helping LGs, including counties and cities, to improve the state's roadway network through a Local Maintenance and Improvement Grant (LMIG) program, which was grown out of the previous Local Assistance Road Program and State-Aid Program. The LMIG program is managed by GDOT's Office of Local Grants (OLG). Under this program, GDOT allocates funds to each county or city using a formula that is based on its population and total roadway mileage. However, a county or city needs to submit a list of projects with a total value of the formula allocation plus the required 10% or 30% match funding according to the Transportation Investment Act (TIA) (*6*). The new program reduces the pavement condition evaluation workforce needed in GDOT. However, each LG needs a means to manage the roadway network in its jurisdiction, that is, a PMS, to inventory roadway conditions, determine treatment methods, and select candidate projects. Though some LGs do periodically collect pavement surface conditions, few of them have a comprehensive PMS to manage their roadway networks effectively within their stringent budgets.

AS well as to fulfill the needs of LMIG, an LG also needs a PMS to practice its obligation of accountability to the public and the legislature. The authors interviewed representatives from six counties and one city, and each of them emphasized the urgency of a PMS for two reasons. First, an LG needs a systematic and objective assessment of the pavements in its jurisdiction and to be able to show it to the public. Thus, the planned pavement maintenance would be more accountable and result in fewer complaints. Second, the Public Works department of an LG needs a PMS to report to the higher management and the legislature on current and future pavement performance based on the existing budget situation and various what-if scenarios under different budgetary conditions. Normally, in a what-if scenario, with a given budget, the overall pavement performance will be predicted. If the prediction is based upon the actual pavement conditions, the pavement deterioration trends, and the best treatment selection criteria, it is much more objective and transparent than a rough estimation. Consequently, it is more likely to convince the legislature and lead to more sufficient pavement maintenance funds.

Challenges of Directly using GDOT's PMS in a Local Government Authority

GDOT has managed and maintained its 18,000-centerline-mile highway system through its established, comprehensive PMS along with the pavement distress protocol, Pavement Condition Evaluation System (PACES) (*7*), since the late 1990s (*8, 9*). LGs in Georgia have sought assistance from GDOT through OLG and LTAP. However, GDOT's full-fledged PMS is not suitable to be directly transferred to and shared with LGs for the following reasons.

First, GDOT's distributed, client-server-based PMS demands significant resources in information technology (IT) infrastructure and technical support, which was originally developed by the Georgia Institute of Technology (Georgia Tech). This type of system was designed for a large organization in which pavement engineers are assigned different roles such as field data collection (GDOT has more than 50 field engineers collecting pavement condition data), data quality checking, data uploading, project selection, and funding allocation, and it needs significant IT support, for example, a dedicated database administrator, and onsite technical support in each of GDOT's seven working districts. In contrast, an LG normally maintains a much smaller pavement network consisting of several hundreds of miles of pavements including both arterial, collector, and residential roads. In an LG's Public Works department, there are normally one or two pavement engineers and IT support is also very limited in comparison with a state DOT. Thus, an in-house-maintained distributed architecture is not suitable for an LG's PMS.

Second, GDOT's roadway network partition is not suitable for an LG. To rate the network-level pavement conditions, the roadway network needs to be partitioned into smaller projects (*10*). The pavement structure, roadway geometry, and pavement conditions should be uniform in each project. However, uniformity is not easy to maintain for a long project, such as a project of more than 10 miles. Thus, each project is further divided into segments. The typical length of a segment is one mile except for the starting and ending ones, which may vary in length. The actual pavement condition survey is performed for each segment and a project-level survey result can be obtained by averaging all segment-level survey results. This type of roadway partition is, however, not suitable for an LG because the basic pavement maintenance unit is much shorter than a project unit in GDOT's practice. For example, a street block of several hundred feet could be a maintenance unit, and so a pavement condition survey needs to be performed for each street block. Also, there is no need for a project-segment partition of its roadway network.

Third, GDOT's LRS cannot be used in an LG. An LRS is normally used to spatially locate a linear feature, such as a roadway section. It defines a point using the measurement from a starting point on a roadway. In PACES, pavement projects and segments are spatially defined by using GDOT's LRS, which uses RCLINK and milepoints. An RCLINK is a 10-digit code defined to identify each roadway in Georgia; it consists of route type (1 digit), county (3 digits), route number (4 digits), and route suffix (2 digits). Though many LGs have GIS maps, few of them has developed an LRS. Therefore, an LRS-based location reference is not suitable in an LG. Instead, a generalized method is needed to define the location of a road or a street block. Based on extensive discussion with several representatives from LGs, a street block was used as the basic unit for pavement condition survey. A street block ID was used for its unique identity and included the starting point and ending point (normally intersecting street names). The street block ID can be uniquely assigned by the software or predefined by an LG.

To address the above challenges, the Office of Research (OR) in GDOT initiated a research project to facilitate the transfer of GDOT's pavement management technology and practice to Georgia's LGs. The Computerized Pavement Condition Evaluation System for Counties and Cities (COPACES-CC) was developed, which leverages some levels of GDOT's existing pavement management practices and available resources as described below. As defined in the NCHRP Report 768 (*11*), three parties—the source (GDOT), the recipients (LGs in Georgia), and the third party (Georgia Tech)—were involved in this T^2 activity. Please note that the NCHRP Report 768 was not directly applied because this research project started before its publication. However, we found that many key factors coincide with the general concept in this report. Thus, some of its terminology was adopted in this paper.

Selection of Pavement Distress Protocols

A pavement distress protocol is indispensable in a PMS, which is used to define the pavement surface distresses, such as cracking, rutting, and raveling. It also defines how to rate the distresses in terms of extent and severity level. A scoring system is then used to calculate the rating of the overall pavement conditions. Popular pavement distress protocols include pavement condition index (PCI) (*4*), pavement surface evaluation and rating (PASER) (*12*), and the present serviceability rating (PSR) (*13*). Pavement condition survey software normally implements one type of pavement distress protocol. For example, PCI was used in MicroPAVER and StreetSaver, and the PASER rating system was used in RoadSoft GIS.

The selection of a pavement distress protocol by an LG should not be arbitrary. Four major factors need to be considered. First, if an LG has already used a protocol for a long time and accumulated sufficient history of pavement condition data, it is a wise choice to keep a similar protocol. A long history of pavement condition data is of unique value to establish accurate models of pavement deterioration. Second, the selected distress protocol should be accurate enough to reflect the actual pavement conditions and facilitate informed decision making on pavement treatment. Normally, a more complicated distress protocol performs better to uncover pavement defects in detail. Thus, a more cost-effective treatment method could be devised. Third, an LG should consider the cost of a network-level pavement condition survey when a pavement distress protocol is selected. Because the complication varies among different protocols, a more complicated distress protocol normally means higher cost for data collection. Finally, an LG needs to consider the existing knowledge and available expertise on pavement maintenance and management.

PCI was developed by the U.S. Army Corps of Engineers and has become an ASTM standard (*4*). In PCI, 19 types of distress are defined along with a complicated rating computation process. GDOT's PACES is not a national standard like the other three protocols, in which 10 types of distress are defined. However, its history can be traced back to the 1970s. GDOT has accumulated about 30 years of well-documented PACES data for its pavement network and PACES has become a common "language" among GDOT's pavement engineers. It is noted that almost every state DOT has its own pavement distress protocol like that of GDOT.

PASER was developed by the University of Wisconsin-Madison Transportation Information Center and is popular among some LGs. Though various types of distress are considered in PASER, the rating (0 to 10) is directly assigned by a pavement engineer based on the observed distress. In the meantime, treatment methods are also directly related to a PASER rating. In PCI and PACES, treatment determination is normally based on overall rating and specific distresses. Thus, they are considered more accurate. PSR was developed at the AASHO Road Test facility (*12*). A PSR rating is assigned by a rater based on the level of service the pavement provides, which is very subjective. PSR is mainly used for the purpose of pavement structural design.

Based on discussions with staff from numerous LGs, it was suggested that GDOT's PACES (*7*) should be used to facilitate the transfer of GDOT's current pavement management technology and practice. The advantages of using GDOT's pavement distress protocol lie in the following aspects. First, GDOT has a long history of successfully using PACES. Many engineers are very familiar with it. Using the same distress protocol, an LG can better utilize the existing resources in GDOT through the LTAP program. Second, GDOT's current maintenance and preservation decision making is based on the PACES rating. Many experienced pavement maintenance engineers have built their knowledge based upon it. By using the same distress protocol, it would be easier for an LG to leverage GDOT's practice in pavement management and maintenance. Third, using PACES will facilitate the communication between Georgia's LGs and GDOT. As mentioned above, an LG is now responsible for providing a list of pavement maintenance projects when it applies for LMIG funds. Using PACES would make it straightforward for GDOT to review the LG's request. Finally, to reduce the cost of training new pavement engineers, Georgia's LGs could hire GDOT's retired pavement engineers. The benefit of this practice is twofold. First, an LG immediately acquires experienced engineers without the need for extensive training. Second, the knowledge from those retired engineers is preserved, which is especially meaningful because knowledge loss has become an issue among the state DOTs.

Nevertheless, GDOT's pavement distress protocol is still somewhat complicated for use in an LG because it requires a rater to walk through the sample location to inspect the extent and severity level of cracks. In considering the popularity and ease of use, PASER is also added in COPACES-CC as a supplement. An LG can use both distress protocols to speed up its pavement condition evaluation process. Currently, two strategies are adopted by the pilot counties. In the first strategy, it is suggested conducting PASER survey first on the LG's entire pavement network. This process is fast because it is a windshield survey method. Then, the pavement projects will be filtered in terms of PASER rating. The pavements with PASER rating less than or equal to 4 (fair condition) will be chosen for further PACES survey. Thus, more informed decisions can be made based on the more detailed PACES rating. In the second strategy, PASER and PACES are used on different pavements in the different functional classes. For residential roads, PASER will be used; PACES will be used on other major arterials and collectors. These two strategies are both valid for LGs and sufficiently utilize the advantages and avoid the disadvantages of the two pavement distress protocols.

System Architecture and Major Functionality of COPACES-CC

Based on the above discussion, Georgia Tech has worked with GDOT to develop a PMS for Georgia's LGs. Figure 1 illustrates the overall system architecture of COPACES-CC, which is a typical cloud-based mobile application that utilizes the most up-to-date technology in the IT industry. COPACES-CC consists of three major components: cloud services, a tablet-based mobile application, and a web application. The following briefly introduces the functions of each component. More technical details can be found in the project's final report (*14*).

The tablet-based mobile application is used by a field engineer to collect pavement condition in the field. It is designed to fully utilize the input convenience of a tablet computer, that is, touch screen instead of keyboard and mouse. When using COPACES-CC for the first time, the inventory of road and segment (i.e., street block) information needs to be downloaded from a cloud database (see "Sync Road" function in Figure 2*a*). The road and segment inventory can be generated manually using the web application or directly loaded to the cloud database. Figure 2*b* shows the pavement condition survey form in COPACES-CC. A rater clicks a button related to a type of distress and inputs the extent and severity level of the distress. After that, the COPACES rating is automatically calculated. In the top-right corner in Figure 2*b*, if "Windshield Survey" is checked, the PASER rating system will be used instead of PACES. As well as pavement condition data, COPACES-CC also utilizes the embedded camera and GPS (if not available, an external GPS dongle is needed) to collect field photographs along with their GPS locations. Field photographs are useful for pavement engineers to diagnose pavement conditions in the office. After field pavement condition data collection is completed, the data along with field photographs can be uploaded to the cloud database by clicking "Upload Survey", as shown in Figure 2*a*.

Cloud computing is currently a major trend in the IT industry. It outsources the major IT-related tasks such as storage, database management system, computing power, development environment, and web services to one or several cloud service providers. The major advantage of using cloud computing is that the end users need not worry about system and data maintenance and can minimize the need for IT support. It can also dramatically reduce the upfront and long-term cost related to IT infrastructure and data management.

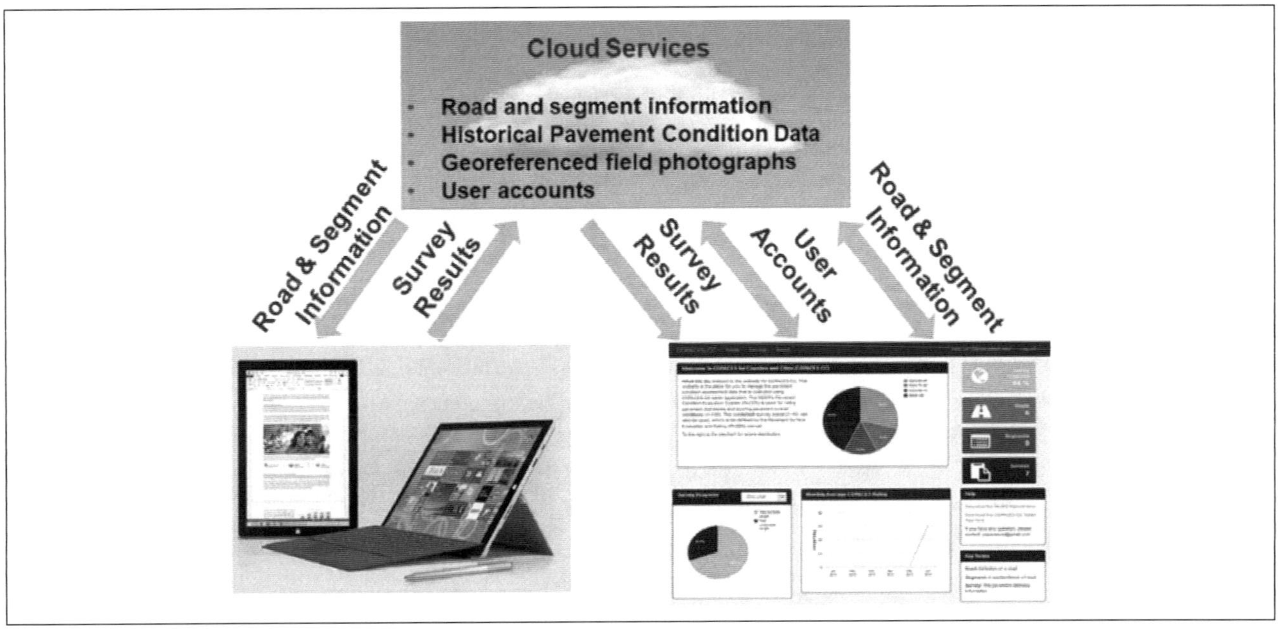

Figure 1. System architecture of COPACES-CC.

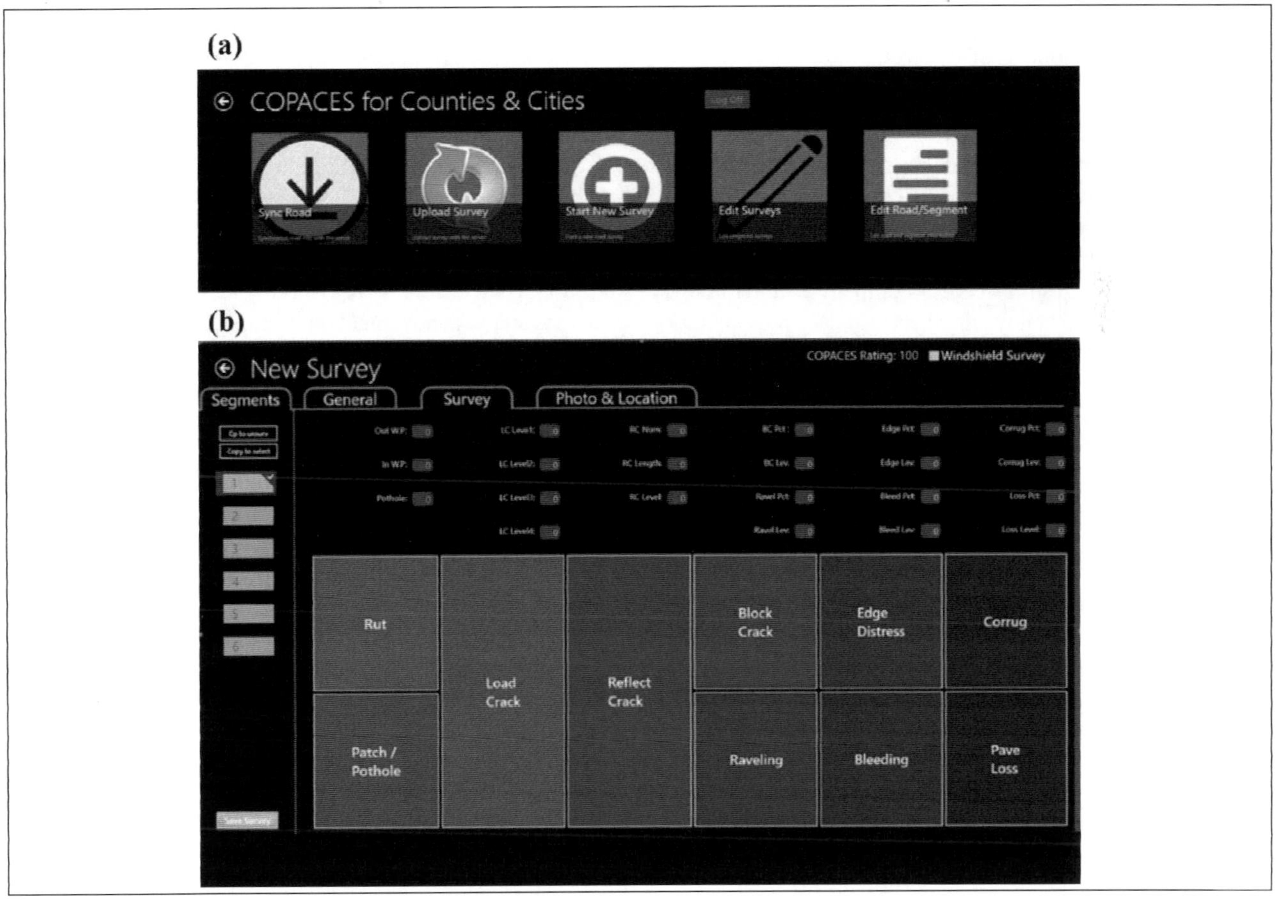

Figure 2. Major functions of COPACES-CC: (*a*) List of functions (*b*) Condition survey form.

Figure 3. Web application of COPACES-CC.

In considering LGs' limitations regarding a full-fledged IT infrastructure, COPACES-CC adopts cloud computing and resides the database management systems, storage, back-end services, and web server in the cloud. An LG can start using COPACES-CC immediately, essentially without any need of investment in IT infrastructure if a compatible tablet computer is available. Currently, cloud services are provided to LGs in Georgia for free. To sustain COPACES-CC in the long run, a small subscription fee might be collected in the future. Nevertheless, the cost of a subscription is very minor in comparison with a major investment in IT infrastructure.

While the tablet-based mobile application is mainly used by a rater to conduct pavement condition surveys in the field, the web application as shown in Figure 3 is a place to manage and report data. The home page of COPACES-CC is designed as a dashboard to show the overall picture of the PMS data. As an example, shown in Figure 3, the top pie chart shows the pavement condition distribution in terms of different ranges of PACES ratings. In GDOT, pavements with PACES rating of less than 70 are classed as in fair condition and might need some treatment. On this dashboard page, other information is also shown, such as the overall progress of the pavement condition survey and the most recent survey results. The "Roads" and "Surveys" pages are related to the road inventory and survey results. End users can manually input roads and segments (i.e., street blocks) using the functions provided by the web application. However, this process could be tedious and time consuming. The research team has worked with several pilot counties and batch-processed their GIS maps to automatically generate the county-wide road and segment inventory. The generated inventory data was then fed into the cloud database. The "Report" page (see Figure 3) provides functions for end users to analyze pavement conditions, suggest treatment methods, allocate maintenance budget, and report data. Given the fact that there is no historical data available for the pilot counties, the treatment determination and budget allocation functions are under development and are not yet released.

Collaboration with GDOT and LGs to Implement COPACES-CC

Given that there are numerous public and private software tools for LGs in the market, the most challenging part for COPACES-CC is not the software itself, but how to leverage GDOT's existing technology and practice and how to get more LGs to buy in. This section summarizes the successful strategies we have used to engage all parties in the development and implementation process of COPACES-CC. Several key T^2 factors coincide with those documented in the NCHRP Report 768 (*12*), though minor differences exist.

At the time of writing, 10 counties had started using COPACES-CC. One county designated COPACES-CC as

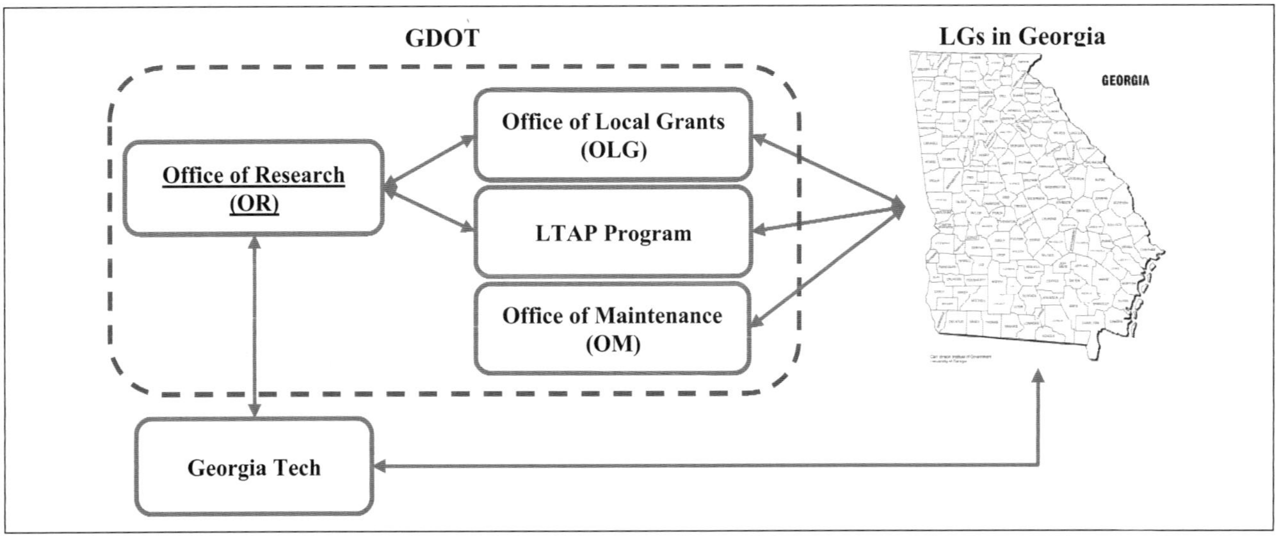

Figure 4. Organizational structure for development and implementation of COPACES-CC.

the required tool for pavement condition data collection and management in its recent Request for Proposal. In the meantime, other counties continuously send inquiries to us about the potential use of COPACES-CC.

Organization of Attending Parties in Technology Transfer of COPACES-CC

Figure 4 illustrates the relationship between all the attending parties. It is noted that the information flow between any two parties is two directional. In GDOT, the direct contacts for LGs are the OLG and LTAP program. LGs also send requests to these two offices for requesting LMIG funds via OLG and asking for technical assistance via LTAP program. The OR aims to initialize research projects, address the department's internal needs and promote technology transfer to the LGs. The Office of Maintenance (OM) serves as a major technology provider in this T^2 activity. OM's responsibility in GDOT is to conduct pavement condition surveys, select maintenance projects, allocate funding, and monitor pavement performance. The pavement engineers in OM are very familiar with PACES and GDOT's pavement management and maintenance practice. In the process of implementing COPACES-CC, OM provides significant technology support in training LGs' pavement engineers on PACES and maintenance practice.

COPACES-CC is a joint project sponsored by OLG, LTAP, and OR to meet the LGs' needs. In the meantime, OM provides significant technology support. During the development and implementation of COPACES-CC, OR also plays a key role to coordinate the activities among all roles, such as the organization of project meetings and workshops, identification of potential end users, and continuous monitoring and support in contract execution. LTAP has direct contacts with all the LGs in Georgia. It provides significant help in managing several statewide workshops. As the developer, Georgia Tech maintains a close working relationship with OR and other offices in GDOT. In the meantime, Georgia Tech also interacts directly with LGs after they are referred by GDOT. The above mechanism has all the parties engaged in each phase of development and implementation, which is presented in the following subsection.

Guided Technology Transfer of COPACES-CC

As mentioned above, in the development and implementation of COPACES-CC, many key factors coincide with the 10 components defined in a guided T^2 (*11*), most of which were included in the three-stage procedures shown in Figure 5.

In Figure 5, three-stage procedures, prototype development, small-scale testing and refinement, and large-scale implementation, were adopted in the T^2 of COPACES-CC. A project kickoff meeting was held to discuss the business requirements. Representatives from several invited LGs and GDOT were involved. Major challenges in leveraging GDOT's existing pavement management and maintenance practice and software tools were discussed and preliminary solutions were determined. After that, a prototype software tool was developed quickly, in less than six months, and two counties and one city started testing the prototype software tool. During the testing, the software tool was continually refined based on feedback. At a later stage, the research team worked with just one dedicated county to carefully refine the software tool. This strategy proved effective and efficient when resources were limited. After a comprehensive test, through GDOT's LTAP program, several statewide workshops were held to train other LGs to use the developed software. The development and implementation procedures shown in Figure 5 were applied to ensure the developed software tool can meet LGs' needs and that GDOT's existing

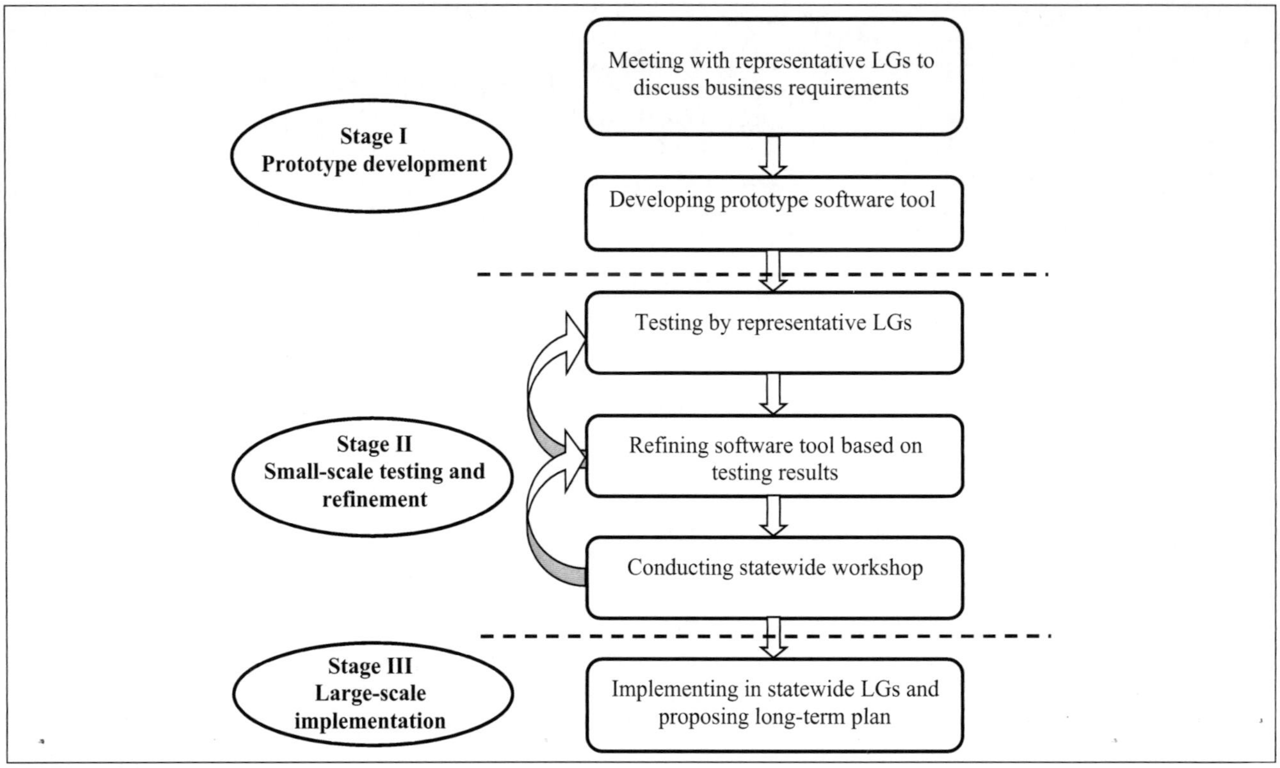

Figure 5. Development and implementation procedures.

practice and resources can be fully leveraged. In addition, the procedures can help mitigate the risk of failure, since each step was sound and required only minimal resources.

The following summarizes some key factors in the T^2 of COPACES-CC, which are closely related to the components defined in the NCHRP Report 768 (*11*).

The success of a T^2 activity largely depends on the identification of an effective and qualified champion in each party. The process of selecting a champion varies from case to case. For example, some champions in LGs were identified through interviews in the statewide workshops. Some others were referred by existing contacts in GDOT and/or other LGs. In the development and implementation of COPACES-CC, the key persons in each party, who are willing to provide enthusiastic and effective support, were heavily involved. For example, in the OR, the Research Implementation Manager actively coordinated the activities of all the roles and helped identify the project funding resources. The directors in the OLG and LTAP of GDOT actively coordinated the communication between Georgia Tech and Georgia's LGs. In each attending LG, key persons of different levels, such as the director of Public Works, the maintenance manager, or pavement engineers, were identified and involved in testing and implementing COPACES-CC.

The engagement of decision makers is another basic factor in a successful T^2 activity. In this case, the project received the support of the higher management of GDOT. When the prototype of COPACES-CC was ready, a demonstration was given to GDOT's Deputy Commissioner. The research team and GDOT's OR, OLG, and LTAP presented the research need and importance of the T^2 activity. In an LG, as well as the champions of different levels, we also arranged meetings with or presentations to the decision makers. This is especially important if the champion in an LG is not at a high level. Direct communication with the decision makers makes the T^2 activity much smoother because the lower-level champion needs less effort to acquire necessary support and decisions can be made more quickly.

Because of the limited funding to support the T^2 of COPACES-CC, only three LGs (i.e., stakeholders) were involved in the development stage (stages I and II, as shown in Figure 5). Thus, the research team could work closely with this small group of users to test, refine, and enhance COPACES-CC. Though the user group is small, their needs reflect most of the other LGs. To engage more LGs to use COPACES-CC in stage III, the research team worked with GDOT's OR and LTAP (Figure 6*a*) to conduct three statewide workshops. Pavement engineers (Figure 6*b*) were invited to introduce GDOT's practice on pavement distress protocol and survey practice. As well as COPACES-CC, the research team also introduced the fundamental concept of transportation asset management and the most up-to-date technology on pavement distress and other asset data collection (Figure 6*c*). These presentations prepared the attendees

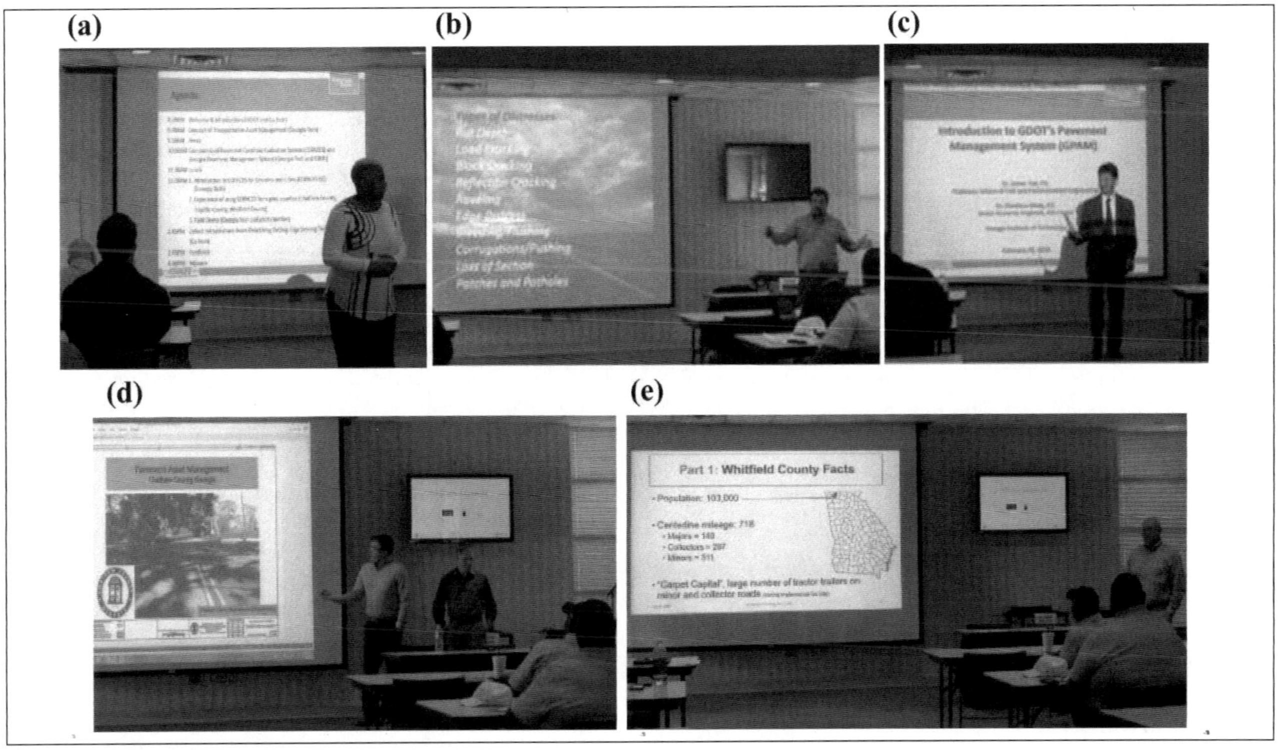

Figure 6. Statewide workshop on COPACES-CC.

with some necessary background knowledge. To demonstrate the use of COPACES-CC in an LG's pavement management practice, the pilot counties were also invited to introduce their experience and known issues (Figure 6*d* and *e*). The engagement of the pilot counties in the T^2 activity of COPACES-CC has proved to be a very effective means to reach out to other LGs. They have become valuable references when new users are thinking of using COPACES-CC.

Long-Term Implementation Considerations

Like all IT products, COPACES-CC needs continuous maintenance, refinement, and enhancement. To be technically and financially viable, GDOT's OR is working with the research team to develop a sustainable T^2 program. This program could be divided into two phases. In the first phase (e.g., 2 years), up to 10 LGs will be chosen to form a user group. Each LG will contribute a small amount of funds to support the refinement and enhancement of COPACES-CC. Other than the pooled-fund raised from these 10 LGs, GDOT considers providing match funding to support the training and operation costs for the new end users. A new user can test COPACES-CC for free for a certain period. After that, it can decide to quit or subscribe as a paid user. In the second phase, GDOT's financial support will be phased out when the user group grows and the subscription fee contributed from all the users can sustainably support the maintenance and operation of COPACES-CC.

Conclusions and Recommendations

To manage the pavement assets in Georgia's LGs cost-effectively, there is a need to leverage GDOT's existing pavement management and maintenance technology and practice. However, due to the organizational difference and resource limitation of LGs, GDOT's technology cannot be directly transferred. To address this issue, the research team has developed COPACES-CC, a pavement management tool based on GDOT's practice, and collaborated with GDOT's OR, LTAP, OLG, and OM to perform the T^2 activity. It is hoped that the successful experience could be useful and beneficial to other states' LGs.

This paper presented the research need and development of COPACES-CC, a guided T^2 process that was adopted in three-stage development and implementation strategies, and future implementation consideration. Lessons learned and future recommendations are as follows:

- A comprehensive survey of statewide LGs is desirable to develop reasonable and sound business requirements before development of the system. However, in this study, only a limited number of representative counties and a city were invited to discuss their business requirements because of the project's tight schedule and limited resources.
- A prototype is needed before comprehensive development. It will help LGs better understand what they can

get and how it works. Based on the testing of the prototype software tool, local agencies can contribute more concrete suggestions and comments, which will mitigate the risk and reduce the uncertainty in the following comprehensive development.

- When resources are limited, using a small group of pilot users is recommended. The pilot users should be qualified in the technical area of interest. Working with a small group of pilot users results in in-depth communication between developers and end users. More concrete refinements can then be made.
- Before the full implementation, successful cases that are built on the pilot users are needed. Their successful experience and testimony will be invaluable to successfully engage new end users.
- In the next step, research will be performed to study the benefits of implementing COPACES-CC in the LGs by quantifying the savings of time and costs.
- To sustain the developed system and make it successful in the long term, a concrete technical and financial plan is needed to support the maintenance, refinement, and enhancement of the system.

Acknowledgments

The authors acknowledge the support provided by Georgia Department of Transportation (Research Project No. 14-27). The authors also thank staff from Newton County, Dougherty County, Moultrie City, Whitfield County, Chatham County, Fayette County, and Dawson County who participated in this project for their technical assistance, suggestions, and comments.

References

1. FHWA. *Asset Management Overview*. Publication FHWA-IF-08-008. FHWA, U.S. Department of Transportation, Washington, D.C., 2007.
2. Wolters, A., K. Zimmerman, K. Schattler, and A. Rietgraf. *Implementing Pavement Management Systems for Local Agencies – State-of-the-Art/State-of-the-Practice Synthesis*. Publication FHWA-ICT-11-094. FHWA, U.S. Department of Transportation, Washington, D.C., 2011.
3. Adjo, A. A., and N. O. Attoh-Okine. Institutional Issues in Implementation of Pavement Management Systems by Local Agencies. *Transportation Research Record: Journal of the Transportation Research Board*, 1996. 1524: 10–15.
4. ASTM International. *Standard Practice for Roads and Parking Lots Pavement Condition Index Surveys*. ASTM D6433-16. ASTM International, West Conshohocken, Pa., 2016.
5. FHWA. *All Public Road Geospatial Representation Study–ARNOLD Reference Manual*. DOT Contract #GS-35F-0001P. FHWA, U.S. Department of Transportation, Washington, D.C., 2014.
6. Georgia Department of Transportation. *Local Maintenance & Improvement Grant (LMIG) Program General Guides & Rules*. Georgia Department of Transportation, 2014.
7. Georgia Department of Transportation. *Manual of Pavement Condition Evaluation System*. Georgia Department of Transportation, Office of Maintenance, 2007.
8. Tsai, Y., and J. S. Lai. Utilization of Information Technology to Enhance Asphalt Pavement Condition Evaluation Program. *The International Journal of Pavement Engineering*, Vol. 2, No. 1, 2001, pp. 17–32.
9. Tsai, Y., and J. S. Lai. A Framework and Strategy for Implementing An IT-based Pavement Management System. *Transportation Research Record: Journal of the Transportation Research Board*, 2002. 1816: 56–64.
10. Haas, R. C. G., and W. R. Hudson, *Modern Pavement Management*. Krieger Publishing Company, Melbourne, Fla., 1994.
11. Hood, M. M., S. R. Thompson, R. J. Vance, M. S. Renz, B. T. Harder, J. Toole, and S. T. Hunter. *NCHRP Report 768: Guide to Accelerating New Technology Adoption through Directed Technology Transfer*. Transportation Research Board of the National Academies, Washington, D.C., 2014.
12. Walker, D. *Pavement Surface Evaluation and Rating Asphalt PASER Manual*. Wisconsin Transportation Information Center, 2002.
13. AASHO. *AASHO Interim Guide for the Design of Rigid and Flexible Pavements*. American Association of State Highway Officials, Washington, D.C. 1961.
14. Tsai, Y., and Z. Wang. *Implementation of an Enhanced COPACES for Georgia's Counties and Cities*. FHWA-GA-16-1427. Georgia Department of Transportation, 2016.

The Standing Committee on Technology Transfer (ABG30) peer-reviewed this paper (18-03591).

Article

TRR
JOURNAL OF THE TRANSPORTATION RESEARCH BOARD

Transportation Research Record
2018, Vol. 2672(13) 29–37

DOI: 10.1177/0361198118794288
journals.sagepub.com/home/trr
SAGE

Reconsidering the Victory Bike in World War II: Federal Transportation Policy, History, and Bicycle Commuting in America

James Longhurst[1]

Abstract
The largest federal intervention in bicycle transportation policy in the 20th century damaged the popularity and prospects of adult cycling in the United States. But in contemporaneous publications and in historical accounts, the World War II "Victory Bike" program has been described positively and fondly, even by bicycle advocates. Using the methodology of the discipline of history, this paper contrasts published literature on the Victory Bike against the unpublished, archival records of the federal government's Revised Ration Order 7 of July, 1942. A first-ever close analysis of month-by-month rationing demonstrates the deeply restrictive nature of that program, which contradicts both early promises and later accounts. By the end of the war, civilian bicycle production and sales had halted completely, the industry had been decimated, and adult cycling was increasingly associated with wartime sacrifice and deprivation. Recovering this 20th century policy history is a necessary part of understanding American bicycle culture in the 21st, partially explaining the comparative lack of adult bicycle commuting today.

Today, cities across the United States are attempting the difficult task of reintroducing adult bicycle transportation to urban streets. Although utilitarian cycling for adults varies between cities and nations, American rates lag behind many other comparable nations (*1*). Why is this so? Though many focus on variance from the global built environment (*2*), the implementation and aftermath of the largest federal policy for bicycle transportation in the 20th century also shaped this outcome.

At the outset of World War II (WWII), the United States instituted an impressively large set of federal programs to efficiently manage strategic resources and train, arm, and deploy military forces while still maintaining a functioning consumer economy. A small piece of that effort was the Revised Ration Order 7 of July, 1942, an action intended to harness the manufacture, distribution, and sale of bicycles to diversify adult transportation options (*3*). Though it was only a small piece of the overall war effort, it still represented the largest federal policy intervention in bicycle transportation in the history of the United States, taking complete control of the bicycle industry. In the decades after the war, in the rare moments it was remembered, this intervention was colloquially referred to as the "Victory Bike" program. The Victory Bike was often described as part of the way that Americans won the war, and later generations of cyclists proudly thought of it as a symbol of the bicycle's true utility in exigent circumstances: a parable of virtue for the edification of noncycling America.

This celebratory history of bicycle transportation in WWII served a purpose for environmentalist and pro-bike rhetoric, and reflected available journalistic sources. But the Victory Bike was not a victory for American bicycling; instead, it served to differentiate American adult bicycling for transportation from global cultures. In a multitude of ways, the Victory Bike program hobbled bicycle manufacturing and severely limited bicycle use rather than promoting it. Its name confuses matters, and tricks our understanding of the meaning of the bicycle's past, present, and future. This paper argues that we should reconsider the program as a defeat rather than a victory. Embracing this interpretation can help us to understand the unique difficulties inherent in bringing the bicycle back to a mono-modal America.

[1]Department of History, University of Wisconsin La Crosse, La Crosse, WI

Corresponding Author:
Address correspondence to James Longhurst: jlonghurst@uwlax.edu

Evidence and Methodology

This study compares a literature survey of the rhetoric of the Victory Bicycle program of WWII from the early war and postwar eras against previously unexamined internal records of the Office of Price Administration (OPA). By comparing the nation's monthly quotas (as well as certificate and production numbers) against this journalistic coverage, the discrepancies between the early 1942 coverage of the Victory Bike program and the actual implementation of Revised Ration Order 7 after July of that year become stark. This first-ever compilation of monthly quotas allows for greater understanding of the grim limitations on wartime cycling, which belies both pre-war boasts and postwar accounts.

Postwar Literature Review

The Victory Bike has become a symbol of hope and success in the 73 years since the war. Mike Davis, writing in the Sierra Club's magazine in 2007, remembers that "[i]n the 1940s, Americans simultaneously battled fascism overseas and waste at home." As a part of that battle, "[m]y parents, their neighbors, and millions of others left cars at home to ride bikes to work." To support that change, recalled Davis, "less than two months after Pearl Harbor, a new secret weapon, the 'victory bike'—made of nonessential metals, with tires from reclaimed rubber—was revealed on front pages and in newsreels" (*4*). Historian Zack Furness describes the positive journalistic coverage of the Victory Bike stories in WWII as a rare instance of the popular press promoting adult cycling "by framing it as a pragmatic and patriotic endeavor meant to bolster the war effort" (*5*). David Herlihy's history of the bicycle contends that "the utilitarian bicycle assumed an even greater presence" with the coming of the war, and that the federal government committed to producing "750,000 bicycles in 1942" to meet wartime needs (*6*). In 2006 the Smithsonian Institution even acquired a Victory Bike, securing the object's place in history. The name "victory bike" has come to be so positive that it has been adopted by shops, manufacturers, and Victory Motorcycles (for some years, a brand of the Polaris Corporation). The name has undergone a transformation in meaning and association—from wartime sacrifice to nostalgic pride.

Wartime Rhetoric

Overview

But before WWII, the idea of putting adults on bikes was one part of the task of reshaping nearly the entire American economy to serve strategic needs. In this total commitment to war, federal agencies planned to intervene not only in consumer choices but in labor agreements, price controls, allocation of resources, and industrial priorities. The Victory Bike program was only a tiny, easily forgettable fragment of a massive project to harness the complete economy for strategic purposes.

The bicycle program was only one among dozens of interventions in the consumer economy, many of which were planned in the tense months (and even years) before America's official entry into the war. In some ways, the OPA and similar domestic wartime agencies were merely extensions of New Deal programs, price controls, and labor agreements. The OPA worked not only to keep down wartime inflation and efficiently distribute consumer goods, but also to enforce limits and prosecute offenders. In the case of the transportation sector, significant limitations on petroleum, rubber, and metals were necessary so that nearly all of those resources would be reallocated for military ships, planes, tanks, and bombs. Only the bare minimum would be diverted to maintain the productive capability of the domestic economy. Transportation policy choices envisioned before the war therefore included severe rationing of gasoline and tires for consumer automobiles, the elimination of new automobiles for civilians, continued transit, and a plan to partially replace automobile travel with adult bicycling.

By January of 1942, just a month after Pearl Harbor, federal agencies announced a rough plan to implement that choice: instead of eliminating the manufacture of new bicycles for the consumer—as had been done with many other consumer goods—the federal government would instead select a small number of manufacturers to build a spartan, standardized design for adults. A large number of new bicycles would be produced each year for military and civilian use, with the civilian portion being prioritized for those with legitimate needs. It was an amazingly detailed, centrally controlled system. It included concentration of manufacturing, standardization on bare-bones designs, price controls, and monthly rebalancing of nationwide stocks.

Early Promises

How would the Depression-weary American public, buffeted by fears of wartime austerity and rapid mobilization, react to news of government-dictated transportation choices? In the first months of the war, the journalistic frame was set with positive stories about the plans for bicycle transportation. Local and national papers alike reprinted accounts of Leon Henderson (soon to be head of the newly designated OPA, previously the Office of Production Management), test-riding prototype Victory Bikes around Washington, D.C.'s national mall in early January of 1942 (Figure 1). The cheerful stories of the publicity stunt seemed a welcome relief for journalists.

Figure 1. Henderson on prototype Victory Bike.
Source: Widely reprinted image of Leon Henderson and Betty Barrett, January 1942. United Press International.

Reproduced in *Life* magazine, newsreels, and countless newspapers, the image of the cigar-chomping Henderson put a positive and patriotic spin on the central message: wartime rationing of gas and tires was coming, but bicycles were one replacement (*7*). Months later, cartoonists could reference the positive imagery of Henderson's ride in otherwise unrelated work, depicting cheerful bankers with onboard stenographers (*8*). Journalistic coverage of the photo opportunity tended to end with a punch line, praising Henderson's "executive ability" as proven by the fact that he could "manage the bicycle, a wide-brimmed hat, a big cigar and a stenographer on the handle bars simultaneously" (*9*).

But the same cheerful stories also foretold the limitations, rationing, and deprivation of wartime. A Geneva, New York account mixed the silliness of the photo opportunity with ominous warnings: "Henderson took time off … to give Mr. and Mrs. America an idea of what wartime transportation may come to in this country." The humorous parts still noted that Henderson "fired up one of his six-inch cigars and took a test run on the new 'Victory' bicycle, carrying pretty Betty Barrett," and that he "managed a few maneuvers in the 'look, no hands' style." But coming sacrifices were in the details—the bike would only be made for adults, with "no copper or nickel," with reclaimed rubber tires "scarcely thicker than a thumb," a far cry from the popular two-inch-wide balloon tires. The bikes would be "[b]uilt plainly, without brightwork or fancy fixtures," out of as little steel as possible and in government-dictated numbers: the newspaper story reported a target of 750,000 new adult bicycles for 1942, in comparison to "1,827,000 bicycles made last year" (*10*).

Even though that yearly target would never be met, the boast of 750,000 new bikes helped set the positive tone for the Victory Bike. A promise that American industry would rise to the occasion and fill in for automobile restrictions, it was repeated throughout early coverage and credulously reprinted in countless histories thereafter. Although 750,000 was a significant restriction compared with previous years, it still represented increased adult bikes, and generously promised new production when other consumer products were entirely curtailed. Not stopping at that already ambitious goal, *Time* magazine reported in January that "manufacturers … agreed that they could produce 1,000,000 bicycles in 1942" if allocated sufficient steel and rubber, and the *New York Times* went even further in March, reporting "a planned output of 2,500,000 for 1943" (*11*, *12*). This would never happen.

Though the general reception of the Victory Bike program was positive, not everyone was pleased. Jack Stinnett called it "a lot of balderdash" in his syndicated column. "America is a land of distances," he argued: "Its problems are not at all those of cycling England, where distances are relatively short." It was "a bit of fallacious reasoning … that the nation is going to take to bicycle wheels overnight." Stinnett specifically called the goal of 2 million bikes per year unrealistic (*13*).

Even more extreme responses appeared. The anti-FDR (and often anti-Semitic) magazine *Social Justice*, a publication of the demagogic Father Coughlin, mocked "Radical Leon Henderson" and rued the coming era of bicycle-borne deprivation and red-tinged collectivization, reporting that Henderson "grunted his satisfaction with this New Deal method of transportation" after dismounting from his ride. The story fretted that "some nationalized factory" would be chosen to build the bikes, further warning of the "possible confiscation of the 10-million bicycles now in operation by school children, messenger boys and others." Bashing wartime plans as authoritarian socialism, the story feared coming austerity: "Here's hoping this damnable war will be over soon … If it lasts from five to ten years, there is the possibility of Americans going around in newspaper undershirts" (*14*).

Even generally-positive wartime coverage understood the Victory Bike program to have the goal of restricting the universe of consumer choices to a limited palette of options. Standardization on adult designs was a wartime sacrifice. In the case of the Victory Bike, the "lightweight" designs were understood to be cheap, utilitarian, unfashionable, and one-size-fits-all.

Thus, at the beginning of 1942, the journalistic frame of the Victory Bike program was set: it was a massive federal investment of strategic resources to design, manufacture and distribute an efficient bicycle to replace missing automobile transportation. Though the rationing, concentration, standardization, and centralized control of this and other OPA programs were understood as sacrifices, journalistic coverage cast them as necessary and even patriotic in the face of wartime conditions. Rationing was intended to control inflation, keep the American homefront functioning, and ameliorate suffering. A stern-faced Leon Henderson argued this point in the newsreel *Washington in Wartime*: "Make no mistake: This regulation is a *war* measure … it is a guarantee to the American people that their cost of living will stay put." Over footage showing Henderson on his prototype bike, the narrator declares that his ride "sets a patriotic example to citizens everywhere" (*15*). Though this early description of the Victory Bike program is often what is remembered, it was not what actually happened.

Wartime Reality

The OPA/War Production Board (WPB) Conflict

In reality, the early promise of extensive production—with targets of 750,000 (or possibly 2 million) new adult bikes per year from 12 different manufacturers—was quickly abandoned, largely without fanfare. Many things changed between the first announcements of the Victory Bike in early 1942 and the implementation of Revised Ration Order 7 in July; even as initial public impressions lasted.

Two historical observations explain this wide divergence. The first is that different agencies were at the wheel; logical assumptions about transportation needs made in 1939—by the National Defense Advisory Commission and the Office of Production Management—were discarded by the War Production Board (WPB) in 1942. As historian Sarah Jo Peterson has pointed out, "in four years, the federal bureaucratic structure shifted four times," which both complicates our later reconstruction of events and explains how official positions might quickly be reversed (*16*).

A second observation can also explain how different agencies could hold diverging points of view on bicycle transportation at exactly the same time. Specifically, two federal agencies engaged in an internal debate in the first year of the war, with very different perspectives on the strategic feasibility of replacing automobiles with bicycles. The WPB was the more powerful of the two, with wide-ranging responsibilities to apportion raw materials, set defense quotas, and harness the economy for strategic purposes (*17*). The OPA was delegated by the WPB to ration what products would be left after military needs were filled. It set prices, divvied up goods, and enforced rules (*18*, *19*).

The wide divergence between agencies—and between pre-war rhetoric and wartime reality—becomes obvious when examining the internal records of the OPA, something that bicycle historians have not previously done. In locations across the United States, the National Archives and Records Administration (NARA) holds the estimated 2,454 items and 12,000 ft^3 of textual, photographic, and film records of the OPA. Under Record Group 188, most of these documents detail the many different consumer rationing programs, from home heating oil to sugar. But some boxes contain records of the Automobile Supply Division of Washington D.C., which managed bicycle rationing. This paper summarizes these previously uncited records from the NARA II at College Park, Maryland, as well as the NARA Regional Archives in Chicago, Illinois.

From these internal records it becomes clear that the OPA and WPB held conflicting views on the viability and desirability of the planned bicycle rationing program. While the OPA stuck with pre-war assumptions that new bicycle (and bicycle tire) production could ameliorate the impacts of gasoline rationing, by June of 1942 the WPB had become deeply pessimistic about the availability of rubber and steel to allow for what it now considered the luxury of bicycle production. For months, the two agencies traded internal reports about which was the more efficient use of strategic resources: new adult bicycles to replace automobile travel, or continued use of the car tires that were already on civilian automobiles (*20*).

Before that conflict broke out, the OPA's Automobile Supply Division assumed that wartime bicycle transportation was desirable, and collected statistics and reports to demonstrate the bicycle's more efficient use of hard-to-get rubber. The Bicycle Manufacturers Association of America wrote to the OPA in April of 1942 in a "Preliminary Statement on Relationship of Bicycle to National Defense." The lobbyists thought the war could alter America's comparative lack of adult bicycle transportation by "forcing the bicycle forward to a position of importance more commensurate with its current development in other countries." From this perspective, auto-centricity was abnormal, and "[n]ational defense exigencies promise to end the motor transport interruption in the normal development of the bicycle in the United States." If automobile travel was to be restricted, "the alternative is individualized transportation with the bicycle. It is cheap in operation. It requires minimum amounts … of essential materials per person transported." A June 15 OPA report concluded that "[i]t would appear that if any new rubber is to be allocated for transportation purposes, it ought to go for bicycles, insofar as bicycles can or will be used, instead of automobile tires."

On the other hand, the WPB wanted to curtail all new bike manufacture. They were making disturbing assumptions about a war that might last a decade or more, and planning for extreme austerity in which citizens would use up their possessions rather than counting on a supply of new or replacement products. The tires that Americans already had on their cars were thus a strategic reserve, and should be used up or recapped before any transportation alternatives were manufactured. The Division of Civilian Supply of the WPB recommended a harsh program that opponents in the OPA summarized in a June 27, 1942 memo as "Railroads – no extra facilities ... Busses – no rubber for new busses . . . Automobile travel [to be] cut 27,000,000 cars to 9,000,000 [or] a cut to 33 1/3 percent of present automobiles. Cut average mileage of remaining cars from 9,000 per year to 5,400 or to 60% ... Stop manufacture of bicycles."

This strict WPB vision of wartime transportation did not win out entirely, but it did temper pre-war plans. Though the Victory Bike program publicly announced in January of 1942 would still continue, the WPB later set the overall production quotas under which it would operate in Limitation Order L-52. Because of the WPB's decision, production would be far smaller than the 750,000-per-year promised in January, when Leon Henderson's ride around the mall made national news. The early rhetoric of the Victory Bike and the later reality of limited availability meant that the federal government had a contradictory approach to adult bicycle transportation: promising millions of new bikes at the beginning of 1942, but slowly decreasing (and eventually eliminating) new bicycle production in the months and years that followed. The public never heard about the OPA/WPB debate, though their experience of wartime transportation would be shaped by it; they would never receive the promised 750,000 new bikes per year.

The first public sign of this change came in April of 1942 – just a few months after Leon Henderson's bike ride and the wave of positive publicity which followed, but still before the OPA/WPB debate over efficiency – when the WPB issued a strict set of emergency regulations, temporarily freezing all bicycle production and sales for civilians. The order was surprisingly draconian; one historian has called it "brutal" (*21*). It was a panicky reaction to perceptions of hoarding in the months following the declaration of war, and declared that "no bicycle may leave a factory, a jobber, a wholesaler, or a retailer's place of business after 11:59 tonight" (*22*).

Over the next few months, a series of agency decisions revealed the newly constrained Victory Bike program, culminating in the order that would set the program for its duration. On May 13, the WPB issued a "supplementary directive" to its April freeze, declaring that new bicycles could be purchased by the military and federal government. It delegated the OPA to control civilian sales of whatever bicycles remained under a further "Rationing Order No. 7" on May 15. All of this was regularized in July under a "full-fledged" and slightly renamed "Revised Ration Order 7" (*3*).

July's revised order followed January's promises in broad strokes, but displayed a fundamental change in approach. Under the WPB's conditions, the order tightened the limitations on bicycle rationing. Notably, the earlier plan for 12 manufacturers was gone; now only two (Huffman and Westfield), would make any new bikes, whether they were for military, Lend–Lease, or civilian use. The cargo-bike model in which Leon Henderson had carried Betty Barret just a few months before was gone too (only one women's and one men's design would be authorized for widespread production). Two separate administrative tools allowed the OPA to selectively limit the distribution of bicycles: a monthly quota of new bicycles available for purchase, set by region and state; and an elaborate system of certificates controlling who could buy the bikes, if indeed any were available (*23*). The reality of wartime control was quickly becoming apparent, even as the rhetoric of cheerful adult bike commuting continued.

Decreasing Monthly Quotas

The internal and inter-agency correspondence preserved in the National Archives demonstrates a restrictive and increasingly austere rationing program, which never produced the promised number of new bicycles. Rather than the planned federal program to produce new and purpose-built Victory Bikes to replace the automobile, the implementation of July 1942's Revised Ration Order 7 more closely resembled the philosophy captured in the Depression-era aphorism of "Use it Up, Wear it Out, Make it Do, or Do Without." Many Americans were denied certificates to purchase new bicycles, the number of available bikes trended steadily downwards, and production for civilian use was completely stopped a full year before the end of the war. Workers would need to rely on public transit and shared cars, buy used bikes, or purchase out of dwindling stocks of already existing bikes.

One sign of this reality can be found in the near-obsessive level of control over new bicycle sales. Throughout most of the 26 months of Revised Rationing Order 7, the Washington OPA office set quotas of

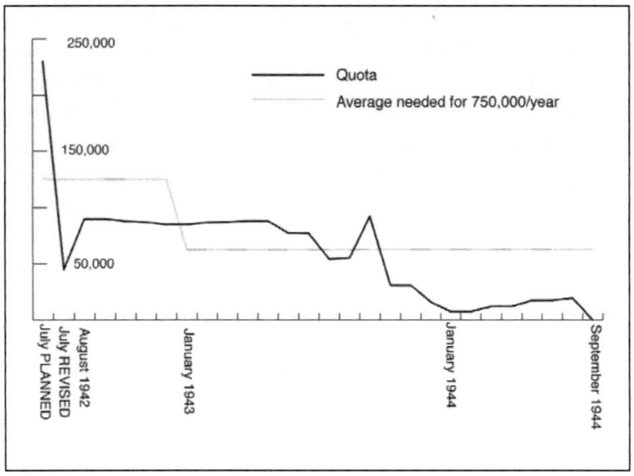

Figure 2. National monthly quota of bicycles for civilian sale compared to the monthly number needed to average 750,000 for the 6 months of 1942 and the 12 months of following years.
Source: National Archives and Records Administration, Record Group 188, College Park, Maryland.

allocable bikes on a monthly and state-by-state basis; sometimes changing total state availability by a single bike from month to month. Amazingly, OPA field offices broke the regional and state allocations down into even smaller amounts, dividing them up between county and city rationing boards, themselves staffed by citizen volunteers who issued certification of wartime need for each and every new bike that was purchased.

The most obvious sign that the Victory Bike was not going to fulfill the optimistic promises of earlier coverage came from the stark reality depicted in Figure 2. During the 26 months of the program, the national quota trended downwards over time, eventually ending at zero in September of 1944. There was nuance here: some declines were small, and in a few instances, quotas were increased by a few percent or reallocated among districts or states. Because of a severe gasoline shortage, quotas were temporarily lifted for the eastern seaboard from June to September of 1943. But overall, there was no possibility that the nation could meet the widely cited promise of 750,000 new bicycles per year.

Throughout the war, the national quota trended downward, with multiple instances of sharp declines. These include the false start of July of 1942, when the first nationally issued quotas were retroactively cut by more than three-quarters immediately after the program began. The *New York Times*, in a story headlined "Output of Bicycles Cut Sharply by WPB," warned ominously that "the reduction announced today sets the quota for July alone at 25 per cent of the amount that had previously been announced as available for July and August together" (*24*). Other cutbacks included November of 1943, which saw quotas reduced to one-third of the previous month; the national total never fully recovered, with only 7,500 new bicycles available for purchase nationwide in both February and March of 1944. This was just months before production of new civilian bicycles ended completely in September of 1944, even though the war continued until September of 1945. In the entire last year of the war, no new adult bicycles were made for civilians, and none remained in reserve. The American bicycle industry was, in effect, zeroed out by the federal government long before the war came to an end.

Other Measurements

Though dramatic, even the declining national quota does not tell the full story of rationing. In effect, it shows the maximum number of bikes that could be sold out of new production and reserves; not the number that was actually sold. A different mechanism of control limited that amount; even if a bike was manufactured and shipped to a retailer in a specific state, it could not be sold to a consumer unless they had a proper wartime need and the certificate to prove it. A July 9, 1942 "Information Bulletin No. 1" outlines these restrictions: certificates were limited to the gainfully employed and war workers who needed to travel at least three days a week, those who had not recently purchased a bike and "those individuals who demonstrate that they will use the bicycle for needed transportation." The complexity of the paperwork was impressive: if the local rationing board was satisfied that a person really needed a new bike, "the Board issues him a Certificate, which is divided into Parts A, B, and C." Part C was proof of need, which stayed with the purchaser. Part A went to the national inventory unit. Part B stayed with the shopkeeper, who would hand collected certificates over to resellers or warehouses when the shop needed new stock; the reseller in turn had to present the collected certificates to manufacturers in order to restock their own inventory. Literally every individual bicycle, and every single transaction, was controlled in this "flow-back" system.

In practice, the number of issued certificates lagged the allocable number of bikes throughout the war. An October 28, 1942 memo addressed this, noting that "Release of bicycles in practically every county have been far below the monthly order … there is some evidence that many local boards have been disinterested in bicycles, have been too busy to give proper attention to bicycle applications" or had been especially punctilious in applying the already restrictive limits on who could receive a certificate. This was partly intentional; it was a technique OPA used across its programs to discourage overconsumption, which one former administrator called "regulation by impedimenta." It was felt that "difficulties of getting a certificate (the right to buy)[;] the belief that they probably could not get one anyway, kept many people from trying to get one." In fact, "[s]ome of the rationing officials invented an aphorism ...'The meek shall inherit the dearth'" (*23*). This lag continued for the duration: "[I]t is especially interesting to note that bicycle certificate issuance has never approached absorbing our quotas," wrote an OPA administrator at the beginning of 1944, while arguing that the program should be discontinued as new production had been largely eliminated. There was no need to ration something that was not available, and for which public demand had been discouraged through the complexities of the certificate process.

The quota was one thing; the number of issued certificates another; and the number of actual new bicycles manufactured yet a third. And that number was alarmingly small. Keeping in mind the early 1942 public prediction of 750,000 bikes per year, the decision by the WPB in Limitation Order L-52 is stark: on October 16, 1942, an internal memo noted that the order "Restricts the manufacture of new bicycles to 10,000 per month." Furthermore, "it is doubtful if the full 10,000 monthly production ticket assigned ... will be met because [the manufacturers] will not secure the required amounts of steel."

This limitation on manufacture explains why the 1943 national monthly quotas were in the tens of thousands, and why the two authorized manufacturers reported a total of only 154,586 new bicycles built for both civilian and military use for *all* of 1943. One internal letter, from March 4 of 1944, reports total production from 1943 for each of the two manufacturing companies broken down by end users. The totals are surprisingly low: 57,755 from the Dayton, Ohio-based Huffman company for the entire year, only 39,068 of which were released by WPB for OPA distribution to civilians. Together, Huffman and Westfield produced 95,913 adult bicycles for civilians in all of 1943—or, a dismal 13% of the 750,000 originally promised. The Victory Bike program, as implemented, barely existed.

Increased Regional Control

A compilation of the monthly rationing quotas by region supports this finding of an increasingly restrictive national quota. Figure 3 presents a first-ever synthesis of monthly press releases collected in the National Archives in College Park, Maryland, with OPA Region III and Region VI records in the Chicago Regional Archives. Though some stretches of time (October, 1943 and February to March 1944) require reconstruction or extrapolation, this analysis demonstrates a continuous downward trend.

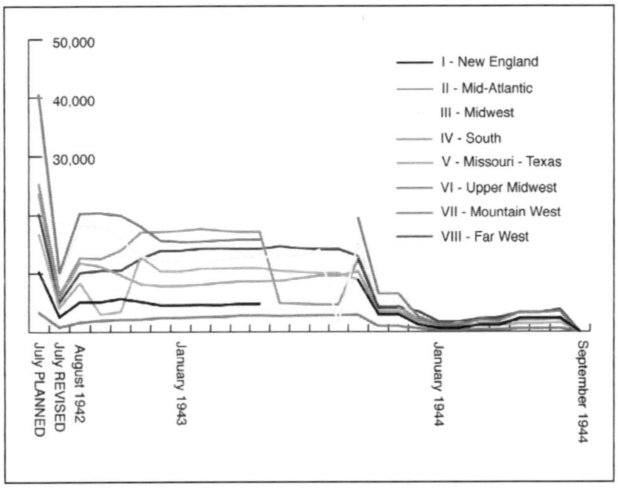

Figure 3. Compilation of OPA's regional monthly quotas of bicycles for civilian sale. Region IX (District of Columbia and territories) excluded.
Source: Reports in NARA II and Chicago Regional Archives.

The regional records also offer a Midwestern case study showing what increasingly restrictive rationing meant on the ground. For February of 1944—with a national quota of only 7,500 bicycles—the seven states of Region VI had just 860 bicycles to divvy up amongst themselves. That month the 127 rationing boards of Cook County, Illinois had just 46 new bikes to distribute among more than four million residents; a preposterous one bike for every 87,000 people.

Discussion

This analysis depicts a much more restrictive bicycle rationing program than recalled in historical memory, with concomitant negative impacts on the bicycle industry and the popular perception of practical riding. Though journalistic coverage in wartime was generally supportive of rationing efforts, family and oral history accounts—many self-published in the last several decades—can fill in missing dissent and complaints about skimpy wartime bikes. One memoir recalls that "[t]he only way to get a bicycle [sic] you had to accept what they called a 'victory bike', that didn't have any more metal than absolutely was necessary" (*25*). Another remembers the Victory Bike as a parental punishment for willfulness in demanding a present: "The bicycle I received on my tenth Christmas was a 'Victory Bike,' The war was going on in that year, and new bikes were basic—no fenders, narrow air tires, sharp banana-style seats, no features. The fact that I got a bike at all during the war was fairly remarkable, but when Dad made up his mind that a lesson would be learned, it was learned" (*26*). Memories of a small town in Oregon lump the

Victory Bike in with gas rationing and hoarding, after which "clothing took the next hit with the introduction of the cuffless, pleatless, lapelless 'victory suit.' The fat-tired, heavy-fendered Schwinn bikes we were so used to were replaced with the stripped-down 'victory bike'" (*27*).

The wartime public's discontent with rationing can also be seen in the letters they wrote to their rationing boards and the OPA. Hundreds of these remain in the National Archives, documenting demands from industry attorneys, complaints from bike shop owners, appeals from schoolchildren, and requests for special treatment. "I'm a man of 60 years old and am helping out the farmers with their work," read a July 8, 1942 appeal from Wawaka, Indiana: "I have no car and have been unable to buy a second-hand bicycle." An OPA State Director wrote on June 23, 1942 to ask for extenuating circumstances for a young man's physical therapy; the OPA assistant General Counsel denied the request, finding that "he is not entitled to acquire a new adult bicycle." From Maryland, an ordained minister asked for a new bicycle; handwritten notes hint at denial. Defense firms had better luck: letters from General Electric and General Motors got immediate responses.

The impact on bicycle manufacturers also resulted in discontent, though wartime dissent was diplomatically phrased. The Schwinn corporation of Chicago, Illinois spent the war producing shells and metal fittings for weapons and aircraft. Still, it appealed—unsuccessfully—to the OPA in 1943 to produce bike repair parts, summarizing an "impartial survey" (emphasis in original) of the 2,450 dealers who were asking for them. The wording was careful to avoid unpatriotic profit-motives: "We know that the necessity for—as well as the desirability of 'keeping 'em rolling', is a decision which rests entirely with the properly appointed officials," it reads. "For our part, we … respectfully urge that we be granted relief to the end that we may supply the cycle repair parts … We urge this since the bike repair shops are vital to the use of cyclists for war time transportation."

By 1945, as the company was preparing for postwar production, Vice President Frank W. Schwinn asked for help. He called the outlook for the bicycle industry "extremely serious." His report estimated that many of the company's existing sales outlets had "closed their doors," before complaining the Victory Bike manufacturer Westfield would have an unfair advantage when consumer sales restarted because it had been "favored" by the government. "Our problems are further complicated by the fact that our losses in sales outlets have assumed alarming proportions and our remaining sales outlets have virtually no stock," Schwinn wrote. "We do not feel that we, and other equally patriotic [firms] … should be abandoned and left helpless in the hands of competitors in the desperate struggle for survival." There is no recorded response. In the coming decades, as Americans turned to auto-centric suburban living and a reduced number of manufacturers focused on the children's market, postwar adult cycling "subsided" in the United States—a reality documented by many observers (*5, 6, 20*).

Conclusion

This comparison of the rhetoric and popular understanding of the Victory Bike program with the internal records of actual production demonstrates conflicts and contradictions. The implementation of strict rationing, combined with limited production, conflicted with the belief that the bicycle would provide essential transportation. This conflict was obvious to government officials even as they proceeded: a July 24, 1942 letter from one OPA administrator noted that "production for civilian uses virtually will cease" under WPB limitations. He ruminated on the strange disconnect between the bicycle's popularity and the WPB's decisions: "In this generation in the United States, this is an unusual situation. The bicycle is being put to new uses hardly to be imagined a year ago … Workers in defense plants and in occupations essential to the community welfare are finding bicycles the most convenient, and in some instances, the only, means of transportation." The whiplash between the idea of needed transportation and limited availability was confusing: "The change has been so sudden and so general, it is difficult to determine just how essential the bicycle is to the war effort."

In the end, production limits made continued rationing pointless. On September 23, 1944, a press release announced that "Rationing of bicycles was ended today." OPA Administrator Chester Bowles explained that "rationing now serves little useful purpose … now the stockpile is gone, and current production is too low to make it worthwhile to burden our War Price and Rationing Boards and the public with rationing." Bowles concluded that "bicycles, which some people expected would play a very important part in our wartime transportation, have actually had a negligible impact on the over-all transportation problem." It's no wonder; there were so few new bikes they could not replace other modes.

This paper argues that the reality of the Victory Bike is too contradictory to be remembered uncritically. While the original goals and conceptualization of the program is attractive for advocates of bicycle transportation's utility and efficiency, the wartime implementation of bicycle rationing was far from the early promises, and most likely limited the ability of adult bicycle commuting to adequately substitute for automobile trips. The most significant federal policy intervention in bicycle transportation in the 20th century did not actually promote adult bicycling: it asked for adults to ride as a wartime sacrifice, but then throttled the supply of the new bicycles needed to support that strategic goal. The history of American bicycling should portray the Victory Bike as a defeat for the viability of adult bicycle transportation, one that contributed to hobbling the feasibility of expanding bicycle mode share in the decades after the war (*20*).

Author Contributions

The author confirms sole contribution to the paper as follows: study conception and design; data collection; analysis and interpretation of results; and draft manuscript preparation. The author reviewed the results and approved the final version of the manuscript.

References

1. Buehler, R., and J. Pucher. International Overview: Cycling Trends in Western Europe, North America, and Australia. In *City Cycling* (Pucher, J., and R. Buehler, eds.), MIT Press, Cambridge, MA, 2012, pp. 9–30.
2. Pucher, J., and R. Buehler. Cycling for Everyone: Lessons from Europe. *Transportation Research Record: Journal of the Transportation Research Board*, 2008. 2074: 58–65.
3. Office of Price Administration. *New Adult Bicycle Ration Regulations*. USGPO, Washington, D.C., 1942. Reprinted in the *Federal Register* at 7 F.R. 5062. hdl.handle.net/2027/uiug.30112059129418?urlappend=%3Bseq=198. Accessed November 15, 2017.
4. Davis, M. Home-Front Ecology: What our Grandparents Can Teach Us About Saving the World. *Sierra Magazine*, July/August 2007.
5. Furness, Z. *One Less Car: Bicycling and the Politics of Automobility*. Temple University Press, Philadelphia, PA, 2010.
6. Herlihy, D. V. *Bicycle: The History*. Yale University Press, New Haven, CT, 2004.
7. Leon, Henderson, Price Boss, is Never Afraid to Stick His Neck Out. *Life*, January 26, 1942, p. 33.
8. McCall, R. And We'll Look Sweet upon the Seat of a Bicycle Built for Two. *Washington Post*, March 29, 1942, p. B2.
9. Editorial. *New York Sun*, January 16, 1942, p. 24.
10. 'Victory' Bicycles Are Planned for Adult Use. *Geneva, New York Daily Times*, January 17, 1942, p. 4.
11. Business & Finance. *Time*, January 12, 1942, p. 68.
12. Liebers, A. More Bicycles to Keep Nation on Wheels. *New York Times*, March 22, 1942, p. 21.
13. Stinnett, J. Washington Daybook. *Prescott Evening Courier*, January 28, 1942, p. 4.
14. Victory Bike to Replace Auto. *Social Justice*, February 9, 1942, p. 13.
15. *Washington in Wartime, Castle Films*. Washington, D.C., 1943.
16. Peterson, S. J. *Planning the Home Front: Building Bombers and Communities at Willow Run*. University of Chicago Press, Chicago, IL, 2013.
17. Bureau of Demobilization. *Industrial Mobilization for War: History of the War Production Board and Predecessor Agencies, 1940-1945*. USGPO, Washington, D.C., 1947.
18. Office of Price Administration. *A Short History of OPA*. Mansfield, H. C. et al. Historical Reports on War Administration. Office of Price Administration, Washington, D.C., 1948.
19. Manning, T. G. *The Office of Price Administration: A World War II Agency of Control*. Henry Holt, New York, NY, 1960.
20. Longhurst, J. L. *Bike Battles: A History of Sharing the American Road*. University of Washington Press, Seattle, WA, 2015.
21. Fitzpatrick, J. *The Bicycle in Wartime: An Illustrated History*. Star Hill Studio, Kilcoy, Queensland, Australia, 2012.
22. Sale of Bicycles to Adults Curbed. *New York Times*, April 3, 1942, p. 1.
23. Thompson, V. A. *The Regulatory Process in OPA Rationing*. King's Crown Press, New York, NY, 1950.
24. Output of Bicycles Cut Sharply by WPB. *New York Times*, July 17, 1942, p. 27.
25. Rozelle, J. *Roots and Branches: Or Growing Up in Maine*. Xlibris Corporation, Bloomington, IN, 2015.
26. Scott, E. W. *Sammy: An Incidental Life of Fortunate Circumstances*. Lulu.com, Morrisville, NC, 2009.
27. Yates, R. Oregon Voices: Klamath Falls Goes to War: A Personal and Newspaper Reminiscence. *Oregon Historical Quarterly*, Vol. 107, No. 3, 2006, pp. 410–423.

The Standing Committee on Transportation History (ABG50) peer-reviewed this paper (18-01429).

Article

TRR
JOURNAL OF THE TRANSPORTATION RESEARCH BOARD

Transportation Research Record
2018, Vol. 2672(13) 38–49

DOI: 10.1177/0361198118798482
journals.sagepub.com/home/trr
SAGE

A Historical Perspective on the AASHTO Guide for the Development of Bicycle Facilities and the Impact of the Vehicular Cycling Movement

William Schultheiss[1], Rebecca L. Sanders[1], and Jennifer Toole[1]

Abstract

This paper draws from a literature review and interviews to demonstrate the impact of advocacy, research, and culture on guidance for design users, bike lanes, and separated (protected) bike lanes in the American Association of State Highway and Transportation Officials' bicycle guides' content from 1974 to present. In the late 1960s and early 1970s, a bicycle renaissance in America resulted in efforts at the local, state, and federal level to encourage bicycling. After Davis, California, became the first community in the United States to build a network of bike lanes, a new brand of bicycle advocacy, vehicular cycling (VC), formed to oppose efforts to separate bicyclists from motorized traffic based on fears of losing the right to use public roads. Via positions of power and strong rhetoric, vehicular cyclists influenced design guidance for decades to come. Through the 1980s, VC philosophy aligned with a federal view that bicyclists freeloaded from the gas tax, resulting in diminished federal support for guidance and related research throughout the decade. However, the passing of the Intermodal Surface Transportation Efficiency Act of 1991 led to increased bicycle networks and renewed interest in bicycle facility research. Although vehicular cyclists continue to oppose roadway designs that separate bicyclists from motorized traffic, research from the last decade demonstrates networks of separated bike lanes improve bicyclist safety and are necessary to meet the needs of the vast majority of the public who want to bicycle but feel unsafe in many traffic contexts.

A transportation design guide does more than just synthesize scientific data and facts, it reflects the cultural values, opinions, and political climate of the era, and the knowledge and biases of the people involved in the development of the guidance. From the 1950s through the 1970s, the transportation profession focused almost entirely on expanding motor vehicle mobility, as evidenced by the completion of a 26,000-mi network of interstates in 1971 after only 15 years of effort. It is within this context of suburban expansion, increasing automobile dependence, and worsening air quality that interest in bicycling for transportation emerged.

However, as more adults began to bicycle in the 1970s, bicycle crashes increased (see Santa Barbara, CA, example in Figure 1) (*1*). People began to demand safety improvements and political support grew to provide them. For example, in 1971, New York Congressman Ed Koch advocated for a new roadway design approach, saying, "The only way to ensure safety for the many thousands of New Yorkers who want to bicycle is to designate official and exclusive bike lanes" (*2*).

As citizens requested bike lanes, transportation officials were put in the uncomfortable position of having no reliable design guidance. To resolve this issue, the U.S. Department of Transportation (USDOT) and some states began to fund bicycle programs and research. This research and experience led to the development of bicycle facility design guidance at the state (notably California) and national level, including the publication of the first edition of the AASHTO *Guide for Bicycle Routes* (Bike Guide) in 1974.

This paper reviews the evolution of the AASHTO Bike Guide from 1974 to 2012 as a case study, relying on Guide content, published research, as well as a review of articles and interviews of key sources to show how advocacy, research, and culture influenced the history of the

[1]Toole Design Group, LLC, Silver Spring, MD

Corresponding Author:
Address correspondence to William Schultheiss:
wschultheiss@tooledesign.com

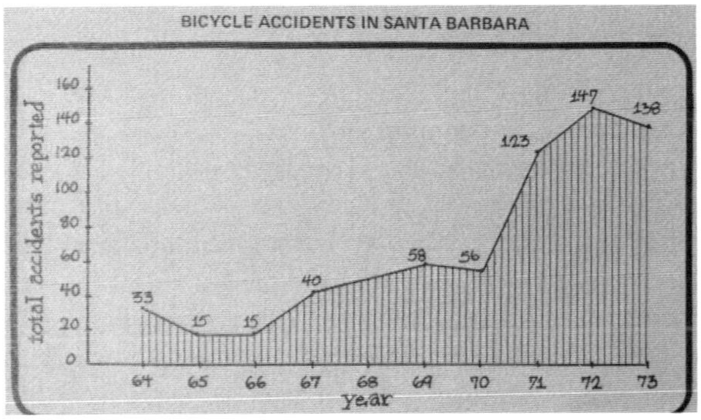

Figure 1. Bicycling trends from 1974 Santa Barbara Bikeway Master Plan (*1*).

Bike Guide as it pertained to guidance for design users, bike lanes (unprotected), and separated bike lanes (curb- or parking-protected).

Methodology

Overview of Methods

As a history paper, this paper focused on a thorough review of the five editions of the AASHTO Bike Guide (from 1974 to 2012), along with foundational research documents from the 1970s to the present that have been sources for the design options in the Guide and, more generally, opposition to and support for bikeways into the present day. To understand the historical context of the various Guide editions, the review also included the published writings of known influential people at the time each edition was written, including papers, books, and web archives. To supplement this information, interviews were conducted with key people involved in the development of research or the various guides who did not have published opinions; these sources were chosen either due to their authorship of the guides, their role in funding the development of the guidance, or their name recognition via multiple sources as a major influence on the process. The interviews served to further illuminate the factors influencing the content of the various guides and provided clarity regarding the impact of advocacy, research, and culture exerted on the final Guide content.

Limitations

As a journal article assessing Guide content and significant relevant influences over time, it was not possible to thoroughly explore all themes that influenced this history. These include themes of power and voice, particularly through the lenses of gender and professional position. Future research further exploring these themes in the context of bicycling history would be welcomed.

Findings

Bicycling Reemerges as a Mode of Transportation: 1960–1974

Davis, California: the First Bike Lane Network in America. In the 1960s and 1970s, Davis emerged as a leader among U.S. cities by providing a bicycle network for its growing population of utilitarian bicyclists commuting to work and the University of California Davis campus. Although conditions for bicycling were favorable on campus, the bicycling experience deteriorated on shared city streets as motorized traffic volumes grew and conflicts between users increased. In 1963, advocates, informed by visits to the Netherlands, began to lobby Davis City Council to support the installation of bike lanes to create a bicycle network that was comfortable and appealing for people of all ages and abilities, including women, children, university students, and seniors (*3*).

Though shared-use paths had existed in the United States for over 60 years, separating bicyclists from motorized traffic within the street was a new concept. In 1966, the City adopted a plan for a network of bike lanes. As bike lanes were new, there was a desire to experiment with a variety of options observed in the Netherlands, including a conventional bike lane (unprotected), a street-level parking-separated bike lane (protected), and a sidewalk-level separated bike lane. By 1972, the City had installed a connected bike lane network that linked the majority of the community (*3*).

Relevant Literature and Research. In 1971, the California legislature mandated a research project to "study the most feasible and least expensive methods by which existing and future public streets and thoroughfares can more safely accommodate bicycle riders." This resulted in the publication of the *1972 UCLA Guidelines* (*4*), which developed bicycle facility design guidance based on interviews with practitioners and a review of U.S. and international guidance and research. European studies of protected bike lanes all showed similar results: overall bicycle safety improved, and bicycle ridership increased with their implementation, although additional treatments were at times necessary to address increased bicycle crashes at some intersections.

Separately, a 1972 study by De Leuw Cather reviewed the safety performance of the Davis bicycle network installed between 1966 and 1971 and provided recommendations for improvement and expansion of the network (*5*). Findings included that:

Bicyclists and motorists preferred streets with bike lanes to those without;

Separation from motor vehicles and pedestrians increased user comfort;

- Few bicycle/motor vehicle crashes occurred in bike lanes;
- Right turn on red was a major source of conflict between bicycles and motor vehicles;
- Sight distance was an issue on the parking-protected bike lanes at intersections and driveways;
- Bicyclists operated at speeds between 7 and 15 mph on flat grades, averaging 10 to 11 mph;
- Debris in protected bike lanes created operational challenges;
- Sidewalk-level protected bike lanes had challenges with pedestrians, trash cans being stored on them, and frequent grade changes at driveways, which discouraged some bicyclists from using them;
- There were some conflicts with pedestrians exiting cars adjacent to parking-protected bike lanes;
- Some bicyclists did not turn left appropriately from bike lanes; and
- Motorists illegally entered or parked in unprotected bike lanes.

Of particular interest for later debates was the parking-separated bike lane on Sycamore Street, which was one block in length and serviced over 550 cyclists/h during peak weekday hours. Despite a lower crash rate/mile than other facilities in the Davis study, as well as clearly high levels of bicycle traffic, this protected bike lane became a poster child for bad design by bikeway opponents due to its seven collisions over a 2-year period. The study recommended improvements for this bike lane, including phase separation with bike signals at a key intersection, additional parking restrictions to mitigate sight distance issues, and the installation of additional protected bike lanes and intersections to manage conflicts and clarify bicycle transition and left-turn expectations. However, the City did not make these improvements. Former city engineer Duane Copley recalled that the protected lanes were not maintained, and at locations with sidewalk-level protected bike lanes or designated sidewalks "at least 50% of bicyclists rode in the street. Bicyclists didn't like traveling with the pedestrians and they didn't like riding up and down driveway ramps and meandering around utilities in the lane." These designs were replaced with on-street bike lanes sometime after 1977 (Copley, personal communication).

The Genesis of Vehicular Cycling as an Advocacy Concept. The implementation of a bicycle network in Davis attracted considerable attention within California and the United States, inspiring other cities to develop bicycle networks. The development of bike lanes and separated bike lanes, in addition to the designation of sidewalks as bike routes, concerned some club bicyclists that they would lose their right to the road and have to bicycle at slower speeds on "inferior" facilities. These fears were exacerbated when local ordinances were passed to require bicyclists to use bikeways, some of which were nothing more than narrow sidewalks. Additionally, a 1944 Uniform Vehicle Code (UVC) provision was popularized: "Wherever a usable path for bicycles has been provided adjacent to a roadway, bicycle riders shall use such path and shall not use the roadway"; this became known as the mandatory sidepath law. UVC also included a provision known as the "ride to the right rule," which requires bicyclists to ride "as close as practicable to the right-hand curb or edge of the roadway"; this became known as the mandatory bike lane law.

In 1972, the City of Palo Alto, California, began to implement a bicycle network by installing unprotected bike lanes and signed bicycle routes, including sidewalk bike routes, and created an ordinance requiring the use of bike lanes. This attracted the attention of John Forester, a local industrial engineer and amateur bicycle racer. Concerned about the mandatory use ordinance and the potential to be required to bicycle on narrow sidewalks with pedestrians, he became involved with the California Statewide Bicycle Committee, which was tasked with developing proposals to modify legislation and create bikeway standards.

Forester believed bike lanes would increase risks associated with turning motorists, motorists opening doors from parked vehicles, and bicyclists turning left, and, most importantly, delegitimize a bicyclist's right to operate on a street. To prove protected bike lanes were dangerous, he rode his bicycle at roadway bicycling speed on a sidewalk designated for bicycle use and attempted to turn left across all lanes of traffic from the sidewalk at this speed. He published his account of this ride as "the one valid test of a sidepath system" that proved sidepath style bikeways were "about 1,000 times more dangerous than riding on the same roads" (*6*). Although this account was clearly anecdotal, Forester used this experience, as well as his position as an engineer, to oppose bikeways and claim sidewalks, sidepaths, and protected bike lanes were dangerous and would increase liability for designers and cities in the event of a crash.

These events inspired him to author a book titled *Effective Cycling*, which centered on a philosophy that "bicyclists fare best when they act as, and are treated as, drivers of motor vehicles" (*7*). The book explains his methods for "driving" his bicycle in a manner similar to a motorized vehicle, a concept known as "vehicular cycling" (VC) that he later popularized through articles in *Bicycling Magazine*.

1974 AASHTO Bike Guide. The 1974 Guide was published by the AASHTO Standing Committee on Engineering

Operations written by unknown authors (*8*). It begins with the safety need for guidance:

> During the past decade, a bicycle renaissance has occurred in the United States. One area of concern in which the impact is already being felt is the growing conflict between bicycles and motor vehicles in the use of streets and highways. One measure of this growth rate is frequency of accidents involving bicyclists. In 1972, there were over 1,100 fatalities involving bicyclists on public highways, an increase of over 100 percent in a decade.

The Guide predated the VC movement and relied heavily upon the 1972 University of California, Los Angeles guidance and De Leuw Cather reports (*4*, *5*).

The Guide suggested criteria to warrant bike lanes, provided information regarding the design and tradeoff considerations for multiple bike lane types (both protected and unprotected), and recommended intersection treatments "to minimize the number of possible conflict points between bicycles, motor vehicles, and pedestrians within the intersection." The Guide recommends unprotected bike lanes as the preferred option based on challenges identified in the 1972 De Leuw Cather study (*5*). Though it recognized "some form of curb or bumper block provides a more positive means of controlling motor vehicle encroachment," it recommended against barriers because of the potential for barriers to be "hazardous to bicycle operation and obstruct maintenance operations."Table 1 (Appendix) shows information on these elements, along with additional key AASHTO Bike Guide content from 1974 through the proposed 2018 Guide to allow comparison between editions.

Use and Abuse of Power: 1975–1981

Relevant Literature and Research. The late 1970s produced literature and research that would influence the field for decades to come. In addition to *Effective Cycling*, Forester authored the *Cycling Traffic Engineering Handbook*, later renamed and reprinted as *Bicycle Transportation*, which described his vision for creating a profession of cycling transportation engineers. The book explained that the underlying reason for providing bikeways was a government and societal effort to get "cyclists out of motorists' way" promoted as a response to "the ignorance and fear of cyclists" operating in motor vehicle traffic. In his opinion, bikeways would do nothing to improve safety and would in many cases worsen safety (*9*).

Forester also wrote that "every facility for promoting cycling should be designed for 30 mph. It if is not, it will not attract the serious cyclist … and hence it will not be an effective part of the transportation system. A facility that is designed only for childlike and incompetent cyclists encourages the 'toy bicycle' attitude and discourages cycling transportation" (*9*). These statements disregarded those who did not share his point of view, and belied reality: research by Lott in 1976 found the addition of bike lanes in Davis attracted people from other routes, and that certain demographics, such as women over the age of 25 and middle, high school, and college students, were much more likely to ride on the street after bike lanes were installed. Forester's reaction was that it was unethical to "appeal to the public superstition that bike lanes make cycling much safer," claiming there was no proof bike lanes were safer (*10*).

Yet other research of the day also did not support the VC premise. The findings from FHWA's 1975 *Safety and Locational Criteria for Bicycle Facilities* report (*11*) were consistent with modern-day research on bicyclists' preferences and safety: bicyclists preferred separation, there were fewer crashes between bicyclists and motorists on streets with bike lanes compared with streets with shared lanes, and facilities that require bicyclists to move in the contra-flow direction resulted in more crashes. Studies by Cross (1977) and Kaplan (1975) of crash types and crash rates on streets with different types of bikeways, respectively, had similar findings (*12*, *13*).

However, in a misrepresentation of the findings, Forester cited these studies as evidence that *all* methods of separation were unsafe, failing to acknowledge the studies had found that streets with bike lanes were safer than streets without. Additionally, the Cross and Kaplan studies did not include separated bike lanes, but Forester used anecdotal references to problems to claim the protected bike lanes "had been tried in Davis and discarded because of a high accident rate," even though the De Leuw Cather study did not support this claim. This led to opposition to their installation in other cities, including "a bitter fight" over their use on San Francisco's Upper Market Street (*14*).

Other bikeway opponents came from a different point of view, as represented by CalTrans engineer Harold Munn. In a 1974 American Society of Civil Engineers (ASCE) paper, Munn argued that efforts to separate bicycles from the normal flow of vehicular traffic were not practical in the 20th century, in which the priority was to accommodate motorized vehicular traffic. He concluded that "the bicyclist will have no choice but to mix with motorized traffic," and that it would be necessary to convince adult cyclists "to operate their bicycles as they do vehicles" (*15*).

Vehicular cyclists would become well-known opponents to bike lanes, protected bike lanes, and shared-use paths in communities across the United States. Many cited Forester and the critical accounts of research he published in books and magazine articles that were easily accessible. Without access to the paper copies of the

1970s studies Forester critiqued (only recently available via the Internet), transportation professionals, public agency staff, and citizens who were unaware of research contradicting Forester's claims had few sources other than his books for bicycle planning. This opposition to bikeways by strong, vocal bicycle advocates led to transportation professionals being confused about what bicyclists wanted, often resulting in agencies taking the path of least resistance, which commonly meant no accommodations for bicyclists on public streets.

1978 CalTrans Bicycle Guide. Under the direction of California Statewide Bicycle Facilities Committee, a CalTrans developed it's first bicycle design guide between 1975 and 1978, publishing it in 1978 (*16*). The work of the committee was heavily influenced by Forester, who was now president of the California Association of Bicycling Organizations. He appointed his colleague, John Finley Scott, as the bicycling representative to the committee to represent the bicyclists' point of view. According to Forester, "Cyclists, again under my leadership, continually opposed the dangerous proposals [developed in the committee], although some cyclists expressed desire for safety bikeways to make cycling popular". Additionally, "in my attempts to redirect the committee to making cycling safer rather than discriminating against cyclists, I wrote probably about half of the paperwork produced by the committee."(*6*) The two primary references for the 1978 CalTrans Guide were Munn's 1974 ASCE paper (*15*) and Forester's *Cycling Traffic Engineering Handbook* (*9*).

This Guide took a remarkably different view of bicycling from the research conducted to date. It codified VC as the primary method for accommodating bicyclists, stating: "An effective program is one that is conducted in recognition of the fact that billions of dollars have been spent on a road system to allow people to travel almost any place they wish. Most of these roads are sufficient to accommodate shared use by bicyclists and motorists, and hence, most bicycle travel has occurred and will continue to occur on that system" (*16*). It also de-emphasized the role bikeways could play to address safety concerns stating: "Many of the common problems are related to improper bicyclist and motorist behavior and can only be corrected through effective education and enforcement programs" and recommended against protected bike lanes and sidepaths.

In contrast to the 1974 AASHTO Guide, the CalTrans Guide eliminated any guidance for specific speeds or volumes that would warrant separation from traffic.

1981 AASHTO Bike Guide. The next version of the AASHTO Guide (1981) was prepared by Richard Lemieux, an engineer who was the FHWA bicycle program manager in the late 1970s (*17*). He did not recall the 1974 AASHTO Guide, saying he "relied upon the 1978 CalTrans Guide as a starting point for the 1981 Guide," as well as a review of the 1975 FHWA *Bicycle Facilities* report (*11*) for additional guidance. Mr Forester was noted for persistently contacting the bicycle researchers and program managers, including Mr. Lemieux during his time at FHWA. When the Guide was put out for public comment, "Mr. Forester provided over half the comments." Mr. Lemieux dutifully considered them all and incorporated many changes into the document as a result (Lemieux, personal communication).

According to Lemieux, the mindset within agency leadership was that accommodating bicycling on the roads was unrealistic. "The people I worked with genuinely believed that highway users pay highway taxes and nobody else has a right to use that money" (*14*). This was underscored when the bicycle program was eliminated in 1983 and Lemieux was reassigned to a pavement design group.

Given its basis, the 1981 Guide has a similar tone and philosophy to the 1978 CalTrans Guide, emphasizing a VC approach with higher-speed bike travel as a design objective, along with education and enforcement support. It also stated Munn's pessimistic view that society would be unwilling to invest in bike networks stating that the construction of a "separated bikeway system, composed of bicycle paths and lanes" would be too expensive and "not provide for the vast majority of bicycle travel" (*17*).

The bike lane warrant from 1974 was replaced by a vague reference to consider traffic volumes and speeds, and protected bike lanes were prohibited, stating, "Bicycle lanes should always be placed between the parking lane and the motor vehicle lanes." This Guide also took a VC approach to intersection design, stating, "Bicycle lanes tend to complicate both bicycle and motor vehicle turning movements at intersections" (*17*).

Quiet on the Front: 1981–1990

Relevant Literature and Research. A lack of funding and fewer bikeways limited research activity in the 1980s. The most cited study from this decade is a 1988 study by Smith and Walsh that evaluated before and after conditions of two newly installed striped bike lanes on a one-way street couplet in Madison, WI, one of which was striped on the left side of the street, and the other of which was striped on the right. The study found a significant increase in crashes associated with the new left-side bike lane in the first year, but not with the right-side bike lane. Moreover, the increase associated with the left-side lane became statistically insignificant after 1 year, which the authors attributed to a familiarity with (and thus changed behavior regarding) the bike lane (*18*).

Table 5. Group B/C bicyclists, urban section, with parking.

average motor vehicle operating speed	average annual daily traffic (ADDT) volume											
	less than 2,000				2,000- 10,000				over 10,000			
	adequate sight distance		inadequate sight distance		adequate sight distance		inadequate sight distance		adequate sight distance		inadequate sight distance	
		truck, bus, rv				truck, bus, rv				truck, bus, rv		
less than 30 mi/h	wc 14	wc 14	wc 14	wc 14	wc 14	wc 14	wc 14	wc 14	bl 5	bl 5	bl 5	bl 5
30-40 mi/h	bl 5	bl 5	bl 5	bl 5	bl 5	bl 6	bl 6	bl 5	bl 6	bl 6	bl 6	bl 6
41-50 mi/h	bl 6	bl 6	bl 6	bl 6	bl 6	bl 6	bl 6	bl 6	bl 6	bl 6	bl 6	bl 6
over 50 mi/h	na	na	na	na	na	na	na	na	na	na	na	na

1 mi/h = 1.61 km/h

Key:* **wc** = wide curb lane** **sh** = shoulder **sl** = shared lane **bl** = bike lane **na** = not acceptable

* see page 11 for definitions.

** **WC** numbers represent "usable widths" of outer lanes, measured from left edge of the parkign space (8 to 10 ft [2.4 to 3.0 m] minimum from the curb face) to the left stripe of the travel lane.

Figure 2. Group B/C (basic/child) bicyclists, urban section, with parking recommendations (Source: [*20*]).

1991 AASHTO Bike Guide. This Guide (*19*) included a new section to extend the prohibition of protected bike lanes and to discourage construction of sidepaths, justified by nine design challenges listed in the 1978 CalTrans *Planning and Design Criteria for Bikeways in California* (*16*). It also discouraged construction of wide sidewalks, as they do "not necessarily add to the safety of sidewalk bicycle travel. Wide sidewalks encourage higher-speed bicycle use and can increase potential for conflicts with motor vehicles at intersections, as well as with pedestrians and fixed objects" (*19*). The studies by Smith and Walsh (*18*) and Cross (*12*) are cited throughout as evidence that drivers would not be looking for contra-flow riders.

A Resurgence of Interest and Activity: 1991–1999

Relevant Literature and Research. The 1990s saw an increase in bicycle research, fueled by federal funding for bicycle and pedestrian facilities resulting from the Intermodal Surface Transportation Efficiency Act of 1991 (ISTEA), which encouraged the development and implementation of bicycle plans.

In 1994, FHWA published the "Selecting Roadway Design Treatments to Accommodate Bicycles" report (*20*). This document was the first to suggest specific bikeway accommodation treatments based on the skill level of the bicyclists and factors such as speed, motor vehicle volumes, presence of parking, and sight distance (Figure 2). This approach, which claimed confident cyclists (Group A) constituted 5% of present cyclists and the B/C group (basic/child) constituted 95% of present cyclists, was vehemently opposed by VC advocates. As noted by Forester, this "policy then assumes that the B/C group will continue to be the large majority for whom the entire system must be designed. In effect, the FHWA advocates dumbing down the cycling traffic system to suit the desires of the least competent possible users." The real purpose of this policy, he said, was "to promote the highway establishment's major cycling interest, its desire to prevent cyclists from delaying motorists" (*9*).

In 1994, Wachtel and Lewiston published a study determining crash probabilities on various facilities in Palo Alto, California (*21*). The study was widely cited by VC advocates for the conclusions that bicyclists were at 1.8 times greater risk of a collision while riding on the sidewalk than riding within the road, and that, regardless of facility type, cyclists were 3.6 times more likely to be involved in a collision if they were riding against traffic as compared to with traffic. The study only included junction-type crashes and did not account for the 26% of crashes that occurred on segments.

However, when Lusk et al. reanalyzed Wachtel and Lewiston's data to include all crashes, they found that the relative risk of sidewalk riding was statistically the same as riding in the road. Moreover, when direction of travel was accounted for, *the relative risk of sidewalk riding dropped to half of the risk of riding in the street* for cyclists riding with traffic (*22*).

Another study, Moritz's 1998 survey of North American bicycle commuters, found that streets with bicycle lanes were clearly the safest with regards to crash rates, as compared to major and minor streets without bicycle facilities (*23*). This study also found higher crash rates on sidewalks and multi-use trails, although it did not distinguish between the two, or separate falls from crashes with motor vehicles. Regardless, both Wachtel and Lewiston's and Moritz's studies (*21*, *23*) were promoted by VC advocates as providing evidence against bikeways (*24*).

In 1997, the first model evaluating real-time perceptions of bicyclists traveling in actual traffic was published (*25*). The study participants represented a cross section of age, gender, experience, and geography, thereby providing a broader assessment of the general public's perceptions than research geared toward bicycling advocacy groups such as the League of American Wheelmen. Furthermore, the study documented for the first time the strong, positive impact of bicycle lanes on comfort ratings ("level of service") while bicycling.

1999 AASHTO Bike Guide. The 1999 AASHTO Bike Guide was the first to be published after ISTEA became law (*26*). The law required each State Department of Transportation to hire a bicycle and pedestrian coordinator. With this dramatic increase in professional staff, this guide received a higher level of review from people throughout the United States than prior versions, resulting in multiple authors and drafts before finalization.

For the first time, the Guide defined different types of bicyclists, using the classification system recommended in the 1994 FHWA report: A (advanced), B (basic), C (child) bicyclist (*20*, *26*). The draft sought to align bikeway type to design user. For example, a "B" cyclist (casual, novice, recreational) would prefer well-defined separation on arterials. An "A" cyclist (confident, experienced) would be comfortable with a wide outside lane. Ultimately, however, the Guide excluded FHWA's recommendation to align bikeway types to design users. The vague language regarding speeds and volumes from previous editions remained, with the caveat that rider preferences should be considered (*26*).

Increasing Interest in Bicycling: 2000–2009

Relevant Literature and Research. The 2000s saw a large increase in bicycle lane installations along with new design guidance published by cities, including the first protected bike lanes in over 40 years (*27*). Research began to focus on interactions between bicyclists and motorists in various street configurations (*28*) and behaviors of bicyclists and motorists in different contexts (*29*), providing empirical evidence of the effectiveness of various facilities. Dill and Carr documented the connection between bike lanes and ridership in cities , showing a positive and significant correlation at the city scale (*30*), while others began to explore barriers to bicycling and preferences for bicycle facilities, and to quantify the desire among the general population to have more opportunities to bicycle (e.g., Dill and Voros, [*31*]).

2012 AASHTO Bike Guide. The 2012 AASHTO Bike Guide was the first guide developed through an NCHRP research process. This edition contained significant new content, and increased references to support guidance included, although the bicycle design user profile still skewed toward the confident cyclist (*32*).

While the 2012 Guide was being written (2007 to 2009), jurisdictions began to install protected bike lanes, green-colored lanes, bicycle signals, and bike boxes following *Manual on Uniform Traffic Control Devices*'s official experimentation process. Due to interest in innovative bikeway design treatments, the first draft of the 2012 Guide included a chapter entitled "Design Issues on the Horizon," which provided some basic information on separated bike lanes, green-colored bike lanes, bike boxes and raised bike lanes. However, the NCHRP panel overseeing the Guide cut the chapter due to the fact these facilities were new and not yet extensively studied. There was also concern that provision of separated bike lanes guidance would conflict with statements that maintained VC philosophy opposing separation.

The A/B/C typology scheme was eliminated in favor of a more nuanced discussion of user skill and comfort to more clearly articulate a person's tolerance for motor vehicle traffic stress based on the 1997 Bicycle Level Of Service research (*25*). For the first time since 1974, a table was added to specify roadway volumes and speed thresholds for different bikeway treatments. The intersection design guidance removed the VC premise that "bicycle lanes tend to complicate both bicycle and motor vehicle turning movements at intersections," which was not supported by evidence.

This Guide was the first to label shared-use paths that were parallel to roadways as "sidepaths," in order to distinguish them from shared-use paths (or trails) located in independent rights-of-way. The Guide also articulated for the first time the importance of integrating sidepath operations with the roadway, stating that the path should have the same priority as the parallel roadway, although the text also maintained extensive discouragement for building sidepaths.

Bicycling as a Mainstream Mode: 2010–Present

Relevant Literature and Research. Recent years have marked a time of significant advances in the understanding of bicyclists' and motorists' preferences and factors that contribute to the safety of bicyclists operating on or near roadways. Importantly, several studies have directly examined the safety of separated bike lanes and found them to be safer than roadways where bicyclists share travel lanes with motor vehicles. For example, multiple studies of separated bike lanes in Canada and the United States by Lusk and her colleagues have found that the overall relative risk of cycling on a protected bike lane was lower than cycling on the reference streets (*22*, *33*). (Lusk et al. were also the first to publish a review of the AASHTO guidance from 1974 to 1999, finding, similar to this study, that the guidelines did not seem to be rigorously research-based [*33*].) A Dutch study by Schepers et al. found that one-way protected bike lanes are significantly safer than a bicycle lane or no facility (*34*). Furthermore, multiple studies by Teschke et al. (*35*) and Harris et al. (*36*), examining bicycle crashes in Toronto and Vancouver found that protected bike lanes were significantly less likely to be associated with a crash than all other facility types.

Separately, studies by Sanders (*37*) and McNeil and colleagues (*38*) (among others) have documented bicyclists' clear preferences for riding on facilities with greater separation from motorists consistent with findings from the 1970s. Dill and McNeil have also worked to characterize people's attitudes toward cycling and roadway design preferences into typologies of traffic tolerance (*39*).

Research has also documented that the provision of a network of separated bike lanes appears to be key to improving safety and ridership. For example, Buehler and Pucher compare cycling rates and injuries to show that bicycling in the United States has been and continues to be much less safe than in peer countries where networks of protected bike lanes exist (*40*). A frequent criticism by VC bikeway opponents is that confounding factors are the source of these safety gains, not the bikeways themselves, and that a safety in numbers effect is therefore unproven (*41*). However, Marques and Hernandez-Herrador evaluated the impact of ridership and bicycle network construction and expansion in Seville, and found evidence of a strong safety in numbers correlation (*42*).

Finally, national and state organizations looking for more advanced guidance have now published their own guidance on how to design separated bike lanes and other bikeway design treatments that are common in other countries, including National Association of City Transportation Officials' *Urban Bikeway Design Guide* and Massachusetts Department of Transportation's *Separated Bike Lane Design Guide* (*43*, *44*).

Conclusion

This paper covers the evolution of AASHTO Bike Guide content from 1974 to 2012, providing insight into the dynamics and context for each edition. At the heart of the story is a fundamental question posed by USDOT Assistant Secretary John Hilten in 1973:

> Should bikeways be designed to accommodate a smaller number of people moving at the maximum rate of speed achievable by the bicyclist over long distances or should they be designed to accommodate the maximum number of people willing to travel for shorter trips?

An inability to directly address this question delayed the development of urban bicycle transportation networks in North America for decades. The public officials and engineering professionals of the 1970s showed creativity and courage in experimenting with new ideas to support safe bicycle transportation, as did AASHTO, which published a design guide based on limited U.S. experience but supplemented by international research and best practices. However, the lack of maintenance for bikeways, a reluctance to fully commit to separated bike lanes with protected intersections, and ill-considered sidewalk bike routes gave credence to vehicular cyclists' concerns and fueled the rise of the VC backlash in the late 1970s. Many of these concerns would worry any frequent cyclist, and should be carefully considered today as new bicycle facilities are being planned and built.

It is also instructive to see how research standards have changed and bicycling knowledge has become more democratic. John Forester had an outsized impact on bicycle planning and facility design during the 1970s and 1980s. However, his tendency to ignore and even belittle the legitimate concerns and preferences held by the general public who did not enjoy bicycling in high volume or high speed mixed traffic, spurred many a researcher to seek more substantiated truths about facility safety and people's preferences.

Today, this research matters more than one person's experience, and there is a stronger commitment at all levels of government to accommodate and encourage bicycling, and to maintain that commitment. It is also now understood that it is possible to build bicycle networks that are safe and appealing to the general population, and that the delivery of these networks can increase bicycling and improve overall safety outcomes. As the bicycle continues to be seen as a key tool to mitigate the damage wrought by decades of car-centric development, that is, congested roadways, polluted air, inequitable transportation options, chronic disease, and loss of life, it is hoped that the field continues to push forward with innovative designs that are rigorously evaluated, providing the best designs and guidance for professionals seeking to plan and build bicycle facilities.

Appendix

Table 1. Comparison of AASHTO Bike Guide Content in the Various Editions

	AASHTO Bike Guide Edition					
	1974	1981	1991	1999	2012	2019 (anticipated)
Pages of guidance	50	36	51	86	201	400 +
Design user description	Commuter, Recreational, Neighborhood	Commuter: Fast, Direct Recreational: Pleasure, Comfort	Utilitarian: Fast, Direct Recreational: Pleasure, Comfort	Advanced (5%): Fast, Direct Basic/Child (95%): Pleasure, Comfort	Experienced/ Confident: Fast, Direct, but appreciate comfort Casual/Less Confident: Majority of population, prefer comfort	Highly Confident (4% to 7% of public) Somewhat Confident (5% to 9% of public) Interested but Concerned (51% to 56% of public)
Minimum design speed	10 mph minimum	20 mph minimum	20 mph minimum	20 mph minimum	18 mph minimum	15 mph minimum for separated bike lanes and shared-use paths
Desirable design speed	15 mph normally	30 mph desirable	30 mph desirable	30 mph desirable	30 mph maximum	18–30 mph depending upon land use and bikeway context
Bicycle lane installation criteria	Vehicle volume >2,000 ADT Bicycle volume >200 bikes/day Vehicle speeds >30 mph	Inexperienced will avoid high volume and speed roads Commuters will ride regardless	Inexperienced will avoid high volume and speed roads Commuters will ride regardless	Inexperienced will avoid high volume and speed roads Commuters will ride regardless	Experienced/ Confident comfortable on most streets, some prefer separation on high volume and speed streets when available Casual/Less Confident prefer separation or low-volume, low-speed streets	Experienced/ Confident comfortable on most streets, some prefer separation on high volume and speed streets when available Casual/Less Confident prefer separation or low-volume, low-speed streets.
	1974	1981	1991	1999	2012	2019 (anticipated)
Bike lane width	4.0 ft min. at curb 5.5 ft min. adj. to parking	4.0 ft min. at curb 5.0 ft min. adj. to parking >5 ft where vehicle speeds >35 mph or substantial % of trucks	4.0 ft min. at curb 5.0 ft min. adj. to parking >5 ft where vehicle speeds >35 mph or substantial % of trucks	4.0 ft min. at curb 5.0 ft min. adj. to parking + 1–2 ft where parking turnover high >5 ft where vehicle speeds >50 mph or substantial % of trucks	4.0 ft min. >5.0 ft where vehicle speeds >35 mph or substantial % of trucks	4.0 ft min. >5.0 ft where vehicle speeds >35 mph or substantial % of trucks

(continued)

Table 1. (continued)

	1974	1981	1991	1999	2012	2019 (anticipated)
Wide curb lanes	Not discussed	Preferred where no bike lanes exist 12-ft minimum 14 ft desirable, benefits bicyclists and motorists	Preferred where no bike lanes exist 12-ft minimum 14 ft desirable, benefits bicyclists and motorists	Preferred where no bike lanes exist 14-ft minimum 15 ft desirable, benefits bicyclists and motorists	Last resort where no bike lanes exist 13-ft minimum 15 ft desirable, benefits bicyclists and motorists	Last resort where no bike lanes exist 13-ft minimum 15 ft desirable, benefits bicyclists and motorists but may result in higher vehicle speeds
Shared roadways	10-ft lanes, light volume traffic 12-ft lanes, heavy volume traffic	No guidance	No guidance	No guidance	Lower speed and volume preferred, use BLOS to assess	Lower speed and volume preferred, use BLOS to assess
Separated bicycle lane guidance	Recommended treatment	Prohibited treatment	Prohibited treatment	Prohibited treatment	Discussed as one-way bicycle only sidepath	Discussed as one-way bicycle only sidepath
Sidepath guidance	Shared use okay Ideal separation from road: min. 20 ft, preferable 30 ft	Bike only preferred Shared use undesirable	Wide curb lanes, bike lanes, and shared lanes preferable to sidepaths due to 9 separate risk factors	Wide curb lanes, bike lanes, and shared lanes preferable to sidepaths due to 9 separate risk factors	Wide curb lanes, bike lanes, and shared lanes preferable to sidepaths due to 14 separate risk factors	Wide curb lanes, bike lanes, and shared lanes preferable to sidepaths due to 14 separate risk factors
Sidepath intersection guidance	Use protected intersection	Preferable: Avoid; No guidance other than provide good sight lines, consider median	Preferable: Avoid; No guidance other than provide good sight lines, consider median	Integrate with intersection Basic guidance for considerations given to manage conflicts	Integrate with intersection Basic guidance for considerations given to manage conflicts	Integrate with intersection Basic guidance for considerations given to manage conflicts

	1974	1981	1991	1999	2012	2019 (anticipated)
Intersection design elements						
Continue bike lanes to intersection	Encouraged	Drop bike lanes at merge areas with right-turn lanes	Drop bike lanes at merge areas with right-turn lanes; option to dash if right-turn volume low	Dash bike lanes at merge areas with right-turn lanes; option to drop	Dash bike lanes at merge areas with right-turn lanes; option to drop	Encouraged
Mark bicycle crossing	Encouraged	No guidance	No guidance	Discouraged, but may be used for complex intersections	May be used for complex intersections	Encouraged
Provide space for 2-stage turn queue box	Encouraged	No guidance	No guidance	No guidance	No guidance	Encouraged
Protected intersection geometry (offset approach)	Encouraged	No guidance	No guidance	No guidance	No guidance	Encouraged

Note: mph = miles per hour; ADT = Average Daily Motor Vehicle Traffic Volume; ft = feet; min. = minimum; adj. = adjacent; BLOS = Bicycle Level of Service.

Acknowledgments

The authors thank Duane Copley of City of Davis, California (former Senior Civil Engineer), and John Fegan, and Richard Lemieux of FHWA (both Bicycle and Pedestrian Program Managers). Special thanks to Mr. Lemieux for providing access to research studies, design guides, and bike master plans from the 1970s.

Author Contributions

The authors confirm contribution to the paper as follows: study conception and design: WS, RS, JT; data collection: WS, RS; analysis and interpretation of results: WS, RS, JT; draft manuscript preparation: WS, RS, JT. All authors reviewed the results and approved the final version of the manuscript.

References

1. City of Santa Barbara Public Works. *The Proposed Bikeway Master Plan.* City of Santa Barbara Public Works, Santa Barbara, CA, 1974.
2. Cranor, D. *Ed Koch, Force Behind Creation & Removal & of First Green Lanes, Passes Away.* People For Bikes. February 05, 2013. http://www.peopleforbikes.org/blog/entry/ed-koch-force-behind-creation-and-removal-of-first-green-lanes-passes-. Accessed July 22, 2017.
3. Buehler, T. *Fifty Years of Bicycle Policy in Davis, California.* Master's thesis. University of California, Davis, CA, 2007.
4. Fisher, G., S. Hulbert, M. R. Ramey, S. Fass, D. Gonzales, A. Horowitz, J. Kefauver, B. Kobritz, W. Lum, M. Millar, C. Nicodemus, A. Ravindranath, D. Stenson, and E. Weller. *Bikeway Planning Criteria and Guidelines.* University of California, Institute of Transportation and Traffic Engineering, School of Engineering and Applied Science, Los Angeles, CA, 1972.
5. De, Leuw, and Company. Cather *City of Davis Bicycle Circulation and Safety Study Volumes 1, 2, and 3.* Davis, CA, 1972.
6. Forester, J. *American Cycling History from the 1940s; As I Remember It.* http://www.johnforester.com/Articles/Social/My%20History.htm. Accessed July 22, 2017.
7. Forester, J. *Effective Cycling.* MIT Press, Cambridge, MA, 2012.
8. American Association of State Highway and Transportation Officials. *Guide for Bicycle Routes.* American Association of State Highway and Transportation Officials, Washington, D.C., 1974.
9. Forester, J. *Cycling Transportation Engineering.* Custom Cycle Fitments, Palo Alto, CA, 1977.
10. Lott, D., T. Tardiff, and D. Lott. Evaluation by Experienced Riders of a New Bicycle Lane in an Established Bikeway System. *Transportation Research Record: Journal of the Transportation Research Board*, 1978. 683: 40–46.
11. Smith, D. T. *Safety and Locational Criteria for Bicycle Facilities.* Publication FHWA-RD-75-112. Federal Highway Administration, U.S. Department of Transportation, Washington, D.C., 1975.
12. Cross, K. D. *Identifying Critical Behavior Leading to Collisions between Bicycles and Motor Vehicles.* Study Presented to the California Statewide Bicycle Committee, 1977. http://www.johnforester.com/Articles/Safety/Cross01.htm. Accessed July 25, 2017.
13. Kaplan, J. *Characteristics of the Regular Adult Bicycle User.* Federal Highway Administration, U.S. Department of Transportation, Washington, D.C., 1975.
14. Epperson, B. D. *Bicycles in American Highway Planning: The Critical Years of Policy Making, 1969-1991.* McFarland & Company, Inc., Jefferson, NC, 2014.
15. Munn, Harold E. , Bicycles and Traffic, Transportation Engineering Journal, ASCE, Vol. 101, No. TE4 Proc. Paper 11729, pp. 753–762, 1975.
16. California Department of Transportation. *Planning and Design Criteria or Bikeways in California.* Federal Highway Administration, California Department of Transportation, Sacramento, CA, 1978.
17. American Association of State Highway and Transportation Officials. *Guide for Development of New Bicycle Facilities.* American Association of State Highway and Transportation Officials, Washington, D.C., 1981.
18. Smith, R Jr, and T. Walsh, T. Safety Impacts of Bicycle Lanes. *Transportation Research Record: Journal of the Transportation Research Board*, 1988. 1168: 49–56.
19. American Association of State Highway and Transportation Officials. *Guide for Development of New Bicycle Facilities.* American Association of State Highway and Transportation Officials, Washington, D.C., 1991.
20. Wilkinson, W. C. A., B. Clarke, and R. Epperson Knoblauch. *Selecting Roadway Design Treatments to Accommodate Bicycles.* Federal Highway Administration, U.S. Department of Transportation, Washington, D.C., 1994.
21. Wachtel, A., and D. Lewiston. Risk Factors for Bicycle-Motor Vehicle Collisions at Intersections. *ITE Journal*, vol. 64, No. 9, 1994, pp. 30–35.
22. Lusk, A. C., P. G. Furth, P. Morency, L. F. Miranda-Moreno, W. C. Willett, and J. T. Dennerlein. Risk of Injury for Bicycling on Cycle Tracks Versus in the Street. *Injury and Prevention*, Vol. 17, No. 2, 2011, pp. 131–135.
23. Moritz, W. Survey of North American Bicycle Commuters: Design and Aggregate Results. *Transportation Research Record: Journal of the Transportation Research Board*, 1998. 1578: 91–101.
24. Allen, J. S. Research… Reviewed or Critiqued. http://www.bikexprt.com/research/index.htm. Accessed July 25, 2017.
25. Landis, B., V. Vattikuti, and M. Brannick. Real-Time Human Perceptions: Toward a Bicycle Level of Service. *Transportation Research Record: Journal of the Transportation Research Board*, 1997. 1578: 119–126.
26. American Association of State Highway and Transportation Officials. *Guide for Development of Bicycle Facilities.* Washington, D.C.: American Association of State Highway and Transportation Officials, 1999. http://www.industrializedcyclist.com/aashto.pdf.
27. Chicago Department of Transportation and Chicagoland Bicycle Federation. *Bike Lane Design Guide.* Chapel Hill: Pedestrian and Bicycle Information Center, Highway Safety Research Center, University of North Carolina-

Chapel Hill, 2002. https://www.cityofchicago.org/content/dam/city/depts/cdot/bicycling/publications/bike_lane_design_guide.pdf.
28. Hunter, W., J. Feaganes, and R. Srinivasan. Conversions of Wide Curb Lanes: The Effect on Bicycle and Motor Vehicle Interactions. *Transportation Research Record: Journal of the Transportation Research Board*, 2005. 1939: 37–44.
29. Van Houten, R. and C. Seiderman. How Pavement Markings Influence Bicycle and Motor Vehicle Positioning -Case Study in Cambridge, Massachusetts. *Transportation Research Record: Journal of the Transportation Research Board*, 2005. 1939:
30. Dill, J., and T. Carr. Bicycle Commuting and Facilities in Major U.S. Cities: If You Build Them, Commuters Will Use Them. *Transportation Research Record: Journal of the Transportation Research Board*, 2003. 1828: 116–123.
31. Dill, J., and K. Voros. Factors Affecting Bicycling Demand: Initial Survey Findings from Portland Region. *Transportation Research Record: Journal of the Transportation Research Board*, 2007. 2031: 9–17.
32. American Association of State Highway and Transportation Officials. *Guide for Development of Bicycle Facilities*. Washington, D.C.: American Association of State Highway and Transportation Officials, 2012. https://nacto.org/wp-content/uploads/2011/03/AASHTO-Guide-for-the-Development-of-Bicycle-Facilities-1999.pdf.
33. Lusk, A. C., P. Morency, L. F. Miranda-Moreno, C. Walter, W. C. Willett, and J. T. Dennerlein. Bicycle Guidelines and Crash Rates on Cycle Tracks in the United States. *American Journal of Public Health*, Vol. 103, No. 7, 2013.
34. Schepers, J. P., P. A. Kroeze, W. Sweers, and J. C. Wust. Road Factors and Bicycle-motor Vehicle Crashes at Unsignalized Priority Intersections. *Accident Analysis and Prevention*, Vol. 43, 2011, pp. 853–861.
35. Teschke, K., M. A. Harris, C. C. Reynolds, M. Winters, S. Babul, M. Chipman, M. D. Cusimano, J. R. Brubacher, G. Hunte, S. M. Friedman, and M. Monro. Route Infrastructure and the Risk of Injuries to Bicyclists: A Case-Crossover Study. *American Journal of Public Health*, Vol. 102, No. 12, 2012, pp. 2336–2343.
36. Harris, M. A., C. C. Reynolds, M. Winters, P. A. Cripton, H. Shen, M. L. Chipman, M. D. Cusimano, S. Babul, J. R. Brubacher, S. M. Friedman, and G. Hunte. Comparing the Effects of Infrastructure on Bicycling Injury at Intersections and Non-intersections Using a Case-Crossover Design. *Injury Prevention*, Vol. 19, 2013, pp. 303–310.
37. Sanders, R. L. We Can All Get Along: The Alignment of Driver and Bicyclist Roadway Design Preferences in the San Francisco Bay Area. *Transportation Research Part A*, Vol., 91, 2016, pp. 120–133.
38. McNeil, N., C. Monsere, and J. Dill. Influence of Bike Lane Buffer Types on Perceived Comfort and Safety of Bicyclists and Potential Bicyclists. *Transportation Research Record: Journal of the Transportation Research Board*, 2015. 2520: 132–142.
39. Dill, J., and N. McNeil. Revisiting the Four Types of Cyclists: Findings from a National Survey. *Transportation Research Record: Journal of the Transportation Research Board*, 2016. 2587: 90–99.
40. Buehler, R., and J. Pucher. Trends in Walking and Cycling Safety: Recent Evidence from High-Income Countries, with a Focus on the United States and Germany. *American Journal of Public Health*, Vol. 107, 2017, pp. 281–287.
41. Allen, JS. Comments on a safety-in-numbers study. John S. Allen's Bicycle Blog. http://john-s-allen.com/blog/?p=669
42. Marques, R., and V. Hernandez-Herrador. On the effect of networks of cycle-tracks on the risk of cycling. The case of Seville. *Accident Analysis and Prevention*, Vol. 102, 2017, pp.181–190.
43. National Association of City Transportation Officials. *Urban Bikeway Design Guide*. New York City: National Association of City Transportation Officials, 2011. http://www.ocpcrpa.org/docs/projects/bikeped/NACTO_Urban_Bikeway_Design_Guide.pdf.
44. Massachusetts Department of Transportation. *Separated Bike Lane Planning and Design Guide*. Boston: Massachusetts Department of Transportation, 2015. https://www.mass.gov/lists/separated-bike-lane-planningdesign-guide.

The Standing Committee on Transportation History (ABG50) peer-reviewed this paper (18-05962).

TRR
JOURNAL OF THE TRANSPORTATION RESEARCH BOARD

Article

Transportation Research Record
2018, Vol. 2672(13) 50–59

DOI: 10.1177/0361198118798728
journals.sagepub.com/home/trr
SAGE

Parking Is for People: A History of Public Parking in Washington, D.C. (and Implications for Transportation Design Today)

Ryan J. Westrom[1], Chris Shaheen[2], and Rebecca Schwartzman[1]

Abstract

The history and etymology of "parking" track a shift in meaning from a military term associated with lining up wagons to a public parklike space adjacent to a roadway to its current meaning as car storage. Parking thus historically had nothing to do with cars, predating their invention, and this unique history continues to evolve in the context of Washington, D.C. This paper seeks to trace this little-known history through historical research, as well as document the various plans that established public parking in the District of Columbia and then track its transition to the typical current meaning as car storage. Tracing the evolution of parking and connecting it to current work promoting green space and walkability could bring about a renaissance of public parking in its previous meaning, highlighting its story, and have implications for the significance and design of public space not just in the District but across the world.

For most English speakers parking has only one meaning: the place where one stores their vehicle. Parking can be off-street or on-street, in garages or driveways, on public or private property, but always, it is simply where you put your car. The search for parking can be stress-inducing and, despite high prices in certain urban settings, rarely a point of pride. There are now lengthy regulations for location, size, and number of parking spaces and rules for their construction. Like cars themselves, however, the term "parking" in this context is only about a century old. In Washington, D.C. (the District of Columbia, or simply, the District) parking is also associated with cars but has a second meaning related to the city's park and open space system. Parking predates the car and has meant something much different and something far more valuable. Prior to being a space for cars, it was understood to be a parklike area of land running along a roadway. The District of Columbia has the most complete and comprehensive system of parking as parkland in North America and perhaps the world. Parking in this sense is thus exemplified in our nation's capital, serving as a reminder of this history and a model for reviving the value of this space. This paper will discuss the etymological origins of parking and its past application in the District as well as opportunities for the present. The evolution of parking in the District will be traced by reviewing the legislation that helped create it, the plans that put it into practice, and the historic context surrounding it as the city grew in population and vehicle ownership. This history will point to current opportunities for the District, as well as places across the globe, to renew design of this public realm portion.

An Etymological History: What is "Parking"?

Questions of how and where to store modes of transportation seem to have existed for as long as people have been traveling. The connection between parking as a green space or as car storage suggests there are significant ties in the origins of parking that point to prior use as a transportation storage area, which evolved into public space. The familiar notion of providing places for vehicles had its foundation in the storing of carriages in stables in Turkey around 1500 BC (*1*). By the first century BC, Roman cities were making off-street spaces for chariots to ease congestion in addition to implementing laws

[1]District of Columbia Department of Transportation, Planning and Sustainability Division, Washington, D.C.
[2]District of Columbia Office of Planning, Washington, D.C.

Corresponding Author:
Address correspondence to Ryan J. Westrom: ryanwestrom@hotmail.com

against having any vehicles on roadways during the day (*2*). The term "parking," however, does not start there but rather originates with the Germanic word "parruk", circa fourth century AD, denoting the fencing surrounding an enclosed space. It was later adopted as "parricus" in Latin, circa eighth century AD, meaning an enclosure or enclosed area. This laid the etymological foundation for our modern English word, park, and by the 17th century, the use of "park" as a noun describing public recreation space was accepted in English (*3*). "Parking," however, appears to have had a broader meaning related to the verb form. Very early usage of "park" as a verb for storing vehicles had a military connection and by the 19th century, and possibly earlier, parking referred to the practice of lining carriages up parallel within an enclosure (*4*). Such usage for military equipment has been documented during the Civil War in the United States. This use may well have arisen as areas used to store artillery overlapped with places described as parks.

An additional alternative meaning of parking—or "to park"—as a verb for planting trees or creating a patch of parkland for people originated during the 19th century (*5*). The Parking Act of 1870 is an early use of the word and the United States Senate clearly meant it also as a noun describing the green space within the right-of-way that was landscaped whether it was down the middle or along the sides of a roadway (*6*). Its use to describe a landscaped area was not limited to the District of Columbia; Laurie Davidson Cox proposed an unrealized parking system for tree parking along sidewalks, plant parking along medians, and car parking at the curbside in 1915 for New York City (*7*). And its understanding in the early 20th century as a green space along sidewalks to plant trees was common enough that it appears in Sinclair Lewis's work *Main Street* (1920) when he describes the exterior of a small home in the fictitious town of Gopher Prairie, MN, as having, "a concrete sidewalk with a *parking* of grass and mud" (*8*). The realization of this meaning of parking, however, appears to have been greatest in the District of Columbia.

Origins of Public Parking in Washington, D.C.

In the context of the District of Columbia, the public parking system defines a little-known yet District-wide system of green space. Its history is unique in its own right and has been implemented through multiple sets of regulations and city plans.

The L'Enfant Plan

The city's parking system was possible because of street widths called for in the L'Enfant Plan, the original city plan designed by Pierre L'Enfant in 1791. The plan established the grand public spaces and avenues in the nation's capital that are recognized around the world. It envisioned magnificent tree-lined streets and avenues, prominent features in the layout of European cities with which L'Enfant was familiar. This plan created streets with minimum widths of 80 ft and included diagonal avenues with widths up to 160 ft (Figure 1). The excessive width of these streets for an 18th century city would eventually become a maintenance issue for the local government, a problem that would be solved by the creation of the city's public parking system in the late 19th century.

The implementation of the L'Enfant Plan was met with limited funding for public works projects, especially one so ambitious, and focused on the most basic issues related to establishing a network of streets. The earliest regulations for public space and street right-of-way were put in place by George Washington the year the plan was approved and focused on limiting how far parts of buildings could extend into the right-of-way and impinge upon public space. Although very rudimentary in their purpose and intent, these regulations demonstrate an early understanding of the need to protect and maintain public space (*10*). The lack of finances dedicated to improving the capital city resulted in dusty, muddy, barely passable roadways and footpaths that epitomized the District in the mid-1800s (*10*). L'Enfant's idyllic vision of tree-lined boulevards were considered for aesthetics and only realized on the city's most prominent thoroughfare, Pennsylvania Avenue between the White House and Capitol Building. Enclosing public space with a fence if the property owner agreed to plant trees is an indication that trees and landscaping were also seen as desirable (*11*).

The L'Enfant Plan's roadway widths and early regulations hint at the issues and concerns that would eventually shape Washington's public parking system. Envisioned in the same light as the new capital, grand and monumental, the idea of improving streets and avenues and establishing the young democracy took shape. This democracy would be challenged and fought over during the Civil War, an event that almost ripped the nation in half. Its physical impact on the District – the result of a significant increase in the city's population, troops, and artillery marching through the city on a daily basis – compounded the problems of maintaining the city's wide rights-of-way. With the preservation of the Union firmly in place after the war, replacing the image of the nation's capital as a dusty village took on greater importance, gave momentum to creating a city as a symbol of a unified country, and led to policies and regulations that would help realize the grand avenues envisioned by L'Enfant.

The Parking Acts

The transition of L'Enfant's dusty avenues to a system of landscaped boulevards lined with public parking

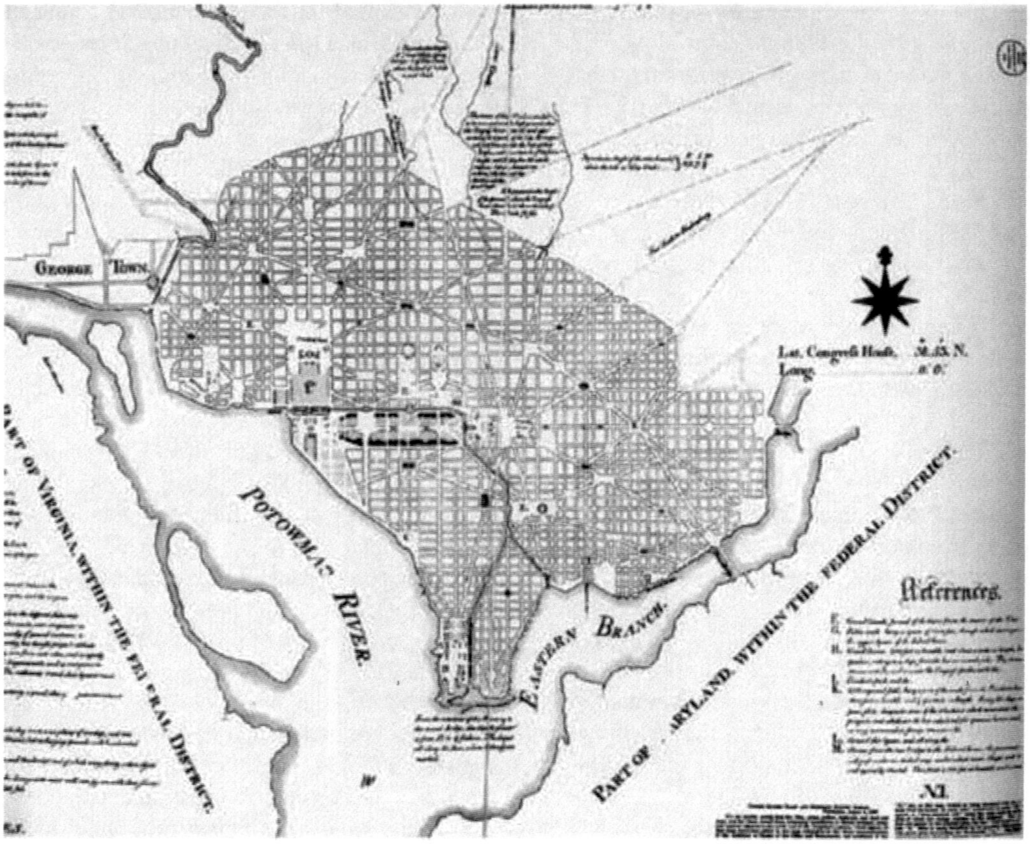

Figure 1. L'Enfant Plan, 1791 (9).

required the government to define and delegate the work needed to accomplish this vision. The legislation that created public parking was initially intended to address practical issues related to maintenance and health, but soon was appreciated for how it transformed the broad avenues and streets of the L'Enfant Plan by designating the space and care it required, as seen in Figure 2. By the mid-19th century, the imperial width of the city's rights-of-way conceived by L'Enfant for a grand capital city not only fell far short of their intention but also constituted a daily inconvenience for the city's residents. Records of the District of Columbia Historic Society document that public parking was conceived by Henry Augustus Willard, member of the city's Board of Public Works during the 1860s and 1870s, as a way to reduce the dust created by the city's wide streets. Willard had a home on K Street NW, one of the widest in the city at nearly 150 ft and home to prominent members of the city including Willard's close associate Mayor Anthony Bowen (*12*). After Bowen complained about dust generated by K Street's unimproved right-of-way, Willard and another associate determined to "park" the roadway, creating a carriage way in the center lined by sidewalks and green space. Local government officials worked with Congress to pass the Parking Act that resulted in the system's implementation throughout all streets and avenues of the city (*12*).

The Congressional Parking Act of 1870 defined public parking as parkland and gave the local government the authority to set aside parts of streets and avenues as parks. This allowed the city to designate space on each street as roadway, sidewalk, and public parking. The area designated as sidewalk was intended to include street trees at the curb, and public parking was defined as the area between the sidewalk and property line envisioned as landscape and greenery. The width of public parking could not exceed half of the width of any street or avenue (*13*). This established a consistent distribution of public space across the city that included a paved portion of street flanked by sidewalks and green space, a pattern that would become and still is a defining image of the District. The Act also led to the creation of the Parking Commission, established in 1871 and charged with its implementation.

Figure 2. Classic parking along street, 1905 (*11*).

A second Parking Act in 1872 went further by defining a system of green and open space comprised on both public and private spaces. The Act established publicly accessible federal reservations – small parks, circles, or squares located where diagonal avenues crossed the traditional grid of streets – and parking that was visually accessible to the public but treated as privately maintained yards. This led to the creation of well-known public spaces like DuPont Circle and Lincoln Park, and established an understanding that public parking was part of a connected system of green space. It also restricted access to public parking to those living in or having business with adjacent properties (*14*). The notion of public parking being the domain of the adjacent property was reinforced by regulations adopted in 1871 that allowed bay window projections into the public parking, a practice that was expanded to include steps, towers, and show windows in 1878. The significance of the 1870 and 1872 Parking Acts was the establishment of a vast network of parkland incorporated into the street right-of-way maintained by both public and private interests that would become clearly understood as part of the city's park and open space system.

As public parking expanded across the city, its value became increasingly appreciated and regulations were adopted to extend it into areas beyond the L'Enfant Plan as land was subdivided for development. Mary Clemmer Ames described this transformation of Washington streets and avenues in 1882: "long vistas of shadeless dust" were now "plat on plat of restful grass [that] '*park*' the city from end to end. Double rows of young trees line these parks as far as the sight can reach. In these June days they fill the air with tender bloom" (*10*). The intent and understanding of public parking as parkland was understood by both government and residents, and remained unchanged throughout the late 19th and early 20th centuries. The Parking Commission oversaw the implementation of the parking system until this responsibility was transferred to the Office of the Superintendent of Trees and Parkings in the late 1890s.

The Highway Plan and Building Restriction Lines

With the benefits of parking firmly understood and desired by residents and city officials, efforts were made to extend this part of the park system into outlying parts of the city as private property was subdivided and developed. The Highway Plan, meant to bring order to streets created by private property owners as they created subdivisions, covered about 60 mi^2 and created a rational and predictable street grid in areas beyond the L'Enfant Plan. The plan, designed by Frederick Law Olmsted, Sr.'s firm in Brookline, Massachusetts, was approved by city officials in phases between 1893 and 1898. Olmsted's influence resulted in a plan with wide avenues that extended from the L'Enfant Plan into what was then the suburbs, connected large parks and open spaces, followed the natural topography, and extended natural areas into residential neighborhoods (*15*). Local officials ensured conformity with the plan by requiring developers to register new subdivisions with the city and obtain approval by the Office of the Surveyor (*15*).

In addition to dictating the layout of the streets, the Highway Plan established minimum street widths of 90 ft to allow a right-of-way width that included public parking. This added additional costs to property owners who now had to turn over a significant amount of their land for public right-of-way. To address this concern, the District allowed property owners to establish a minimum street width of 60 ft if building restriction lines on private property created the effect of a 90 ft right-of-way width. The Parking Act of 1899 gave the city the authority to establish these building restriction lines, which defined areas of private property to be regulated as public parking and extended the parking system to streets and avenues being created in outlying areas of the city (*16*, *17*). Building restriction lines designated an area of private property along a roadway that was regulated as public parking and where the city has the authority to place utility lines or sidewalks. With the explicit goal of extending the parking system to the suburbs, building restriction lines effectively applied parking regulations to private property.

The McMillan Plan

Shortly after adoption of the Highway Plan, the McMillan Plan (1901 to 1902) was developed to extend Washington's system of large and traditional parks into the fabric of the city. Named after James McMillan, a Senator from Michigan and Chairman of the Senate Park Commission, the plan was similarly influenced by a well-known landscape architect of the time, Frederick Law Olmsted, Jr. Its two main components were to establish and restore the monumental core as envisioned by Pierre L'Enfant and expand the traditional park system into outlying areas of the city (*18*). The latter would be realized through large preserves of wooded areas connected by avenues radiating out from the L'Enfant Plan and parkways following ridgelines or valleys. Some of the parkways in the McMillan Plan were first proposed as part of the Highway Plan. Both Olmsted Sr. and Olmsted Jr. understood the value of public parking as part of the city's park system and used it to transform streets and avenues into park-lined boulevards connecting large parks, the most prominent including medians also referred to as parking.

Although some of the parkland proposed in the plan was acquired and used as parks consistent with the plan's recommendations, Congress did not allocate enough funding for its full implementation (*19*). The National Capital Park Commission was created in 1924 to implement the McMillan Plan, which included a coordinated planning effort to improve access to recreation in the city by working with agencies that controlled playgrounds and athletic fields (*19*). Public parking was an important element for realizing the vision of the McMillan Plan by providing the broad streets, avenues, and parkways that tied together the complementary large and natural park areas that are clearly understood as public parks.

Public Space Committee (1935)

By the early 20th century, regulating public space had grown to include more than public parking and become increasingly complicated. There is scant record of the administration of regulating public space in the District at this time, but the establishment of the Public Space Committee in 1935 indicates that requests for private use of public space had increased and required additional oversight. The Annual Report of the Board of Commissioners, a three-member Commission that ran the District's local government, describing the work of the Committee and public space regulations suggests more oversight was needed to review permit requests for the use and occupancy of public space for private purposes. Since it was established, there have been five members on the Committee that represented different agencies within D.C. government. The agencies represented on the Committee have changed over time in response to evolving uses of public space. The Committee continues to govern usage of public parking today via review of public space permit applications for any public parking use, physical changes, or occupancy.

Rapid Evolution in Meaning: A Battle for Parking

The prominence and development of the District of Columbia's parking system in the late 19th century and increasing car ownership in the United States during the late 19th and early 20th centuries makes the city an ideal location to study the use of the term parking and its evolution. The transition of the word "parking" from meaning military storage area or parkland in the late 19th century to its association with storing automobiles in the 20th century is documented by its use in the *Evening Star*, one of the city's leading newspapers. The word parking does not appear in the paper between 1855 and 1869, but after the Parking Act of 1870 it is frequently used to refer to parkland and occasionally in other contexts. Between 1870 to the end of the 19th century, the *Evening Star* commonly reported on public parking as parkland, where it was being installed, meetings of the Parking Commission, or advertisements for properties made more attractive by parking along a street or avenue. This space was intended for use by people,

Figure 3. Horse-drawn carriages and a car find shade along the parking, 1915 (*25*).

especially adjacent residents, and to provide added aesthetic quality to the District's street corridors.

Using the word "parking" for something other than green space was uncommon in the press, but its ubiquitous use in the District and familiar understanding as green space along a street was an easy reference when describing other uses lining streets (although exposure to its usage in the Civil War certainly could have also familiarized people with the term). In 1873, the *Evening Star* reported that "parking of wagons" at street corners or in front of federal reservations was no longer allowed (*20*). In 1875, the newspaper reported on the "parking of carriages" around the Capitol building for the funeral of Vice President Henry Wilson (*21*). The next year, the newspaper reported that B Street, NW, was widened between 7th and 12th streets specifically for the purpose of "parking" wagons in the vicinity of market houses (*22*). In the 1890s, storing railcars for the Baltimore and Ohio Railroad and the District's streetcar company, Capital Traction, became an issue when they "parked" their cars on tracks in the middle of the street (*23*). The practice of storing horses or carriages along or in the landscaped parking evolved into the description for parking vehicles.

This transition of the term to mean storage for automobiles was strengthened by the occasional tying of horses or temporary storage of carriages during special events not just along, but in the parking area (*5*). During the hot summers in the District of Columbia, trees provided shade for horses while their owners were conducting business, in a shop, or visiting a friend. Owners would tie their horses and carriages up to street trees, as shown in Figure 3, sometimes effectively decreasing the width of a street to two travel lanes and two stopped lanes. Although it became illegal in 1872 to trespass on parking, or to cut, injure, or maim parking trees in any way, the convenience and shade provided by the trees for the waiting carriages and horses outweighed the fine levied. Cars were stored alongside horses and as they became ubiquitous were stored in or along the landscaped parking (*24*). The modern-day dilemma of where to park was borne out of the saturation of cars competing for spaces along the street curb.

Systems for controlling traffic and demand for vehicle parking multiplied as car ownership in the District rapidly increased during the early 20th century. In 1907, when standard license plates were adopted, there were 2,200 cars registered in the District of Columbia (*26*). Vehicle registration increased to 112,000 by 1925 when the first traffic light was installed and the Office of the Director of Traffic created. By 1928, registrations had grown to 148,000 and white lines were first painted on streets to separate traffic. In 1931, registrations topped 187,000 and the first traffic-triggered automatic signal control was introduced.

With increasing auto ownership, vehicle parking became an increasingly important issue for local government officials in the city. The first mention of vehicle parking problems was in the Annual Report of the Board of Commissioners sent to Congress in 1924. The report stated that, "parking facilities do not exist and can not [*sic*] be made to exist to the extent necessary to accommodate even a small portion of the automobiles whose drivers seek such facilities. The commissioners have in view, however, a comprehensive program of widening various streets in the congested section of the city" (*27*).

In the District, the car's impact on the green and landscaped parking had not been planned for and became a major issue. By the 1920s, "city officials began cutting down street trees and widening streets to accommodate the volume of cars, thereby [both literally and figuratively] replacing the original meaning of parking as a place for trees and greenery with parking as a place for automobiles to stop" (*28*). Some of the earliest instances of this shift appear in the *Washington Post*, another local newspaper, where the term "parking" was used to explain where cars were parked rather than where trees were "planted" (*28*).

Storing and protecting cars initially designed without a roof, that were difficult to maintain, and required protection from the elements became a necessity and 20th century issue for many North American cities. In 1917, the first recorded "car park" in the United States of America was an upholstery shop at a prime location in Detroit looking to make some extra money (*29*). In the American context, there was a multitude of terms used throughout the automobile's history that referred to the storage of cars. To "rank" a vehicle was synonymous with what we now would call parking (*4*). "Standing" cars referred to vehicles with a driver present that was dropping off or picking up people or goods (*7*). There was also "live" and "dead" parking, the former meaning a standing vehicle that was not transferring passengers or goods whereas the latter was a vehicle that had been dropped off and had no driver present (*4*). Some early terms are still used today, including "cruising" to describe when a driver was in search of a place to leave a car (*7*). By 1927, however, "parking" was generally the accepted term for a car storage area (*5*).

As car ownership increased during the 1920s, managing cars in an urban environment became equal to and sometimes more important than issues related to pedestrians, horse-drawn carts, and other users of the roadway. It is during this time that parking policies supported by the District of Columbia government shifted from parking as green space to making more space for cars. Like other American cities, streets began to take shape around the needs of automobiles and "the public space of the street belonged in law and deed to vehicles" (*30*). This transformation has its roots in the growth of an automobile lobby pushing for more rights for automobiles and the people driving them by framing the uninhibited use of a car as the treasured American value of freedom. Peter Norton observes, "If you ask people today what a street is for, they will say cars" (*30*).

Public Parking as We Know it Today

Today, very little reference to parking in its previous sense as established in the 19th century remains. Even in the District of Columbia, many people are unaware of public parking and colloquially "parking" is almost completely first thought of as a word for vehicle storage. The earlier meaning of parking remains both within government regulations and is still known among some District of Columbia residents. Interestingly, the 2003 Harvard Dialect Survey found that this past awareness exists beyond the District and 1.75% of people surveyed considered parking to mean, "the area of grass between the sidewalk and the road" (*31*). Clusters of such use beyond the District were found in New Jersey and Iowa (Figure 4). The authors are unaware, however, of any other place where a public parking system such as the District's was established and further research is needed to determine whether other communities maintain traditional parking as green space.

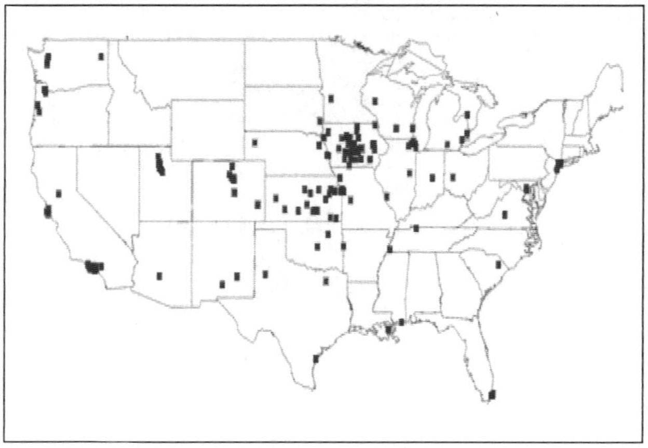

Figure 4. Use of the term "parking" as grassy area, 2003 (*31*).

Contemporary Washington, D.C. Public Parking Regulations

The District of Columbia's public parking system remains a significant public asset regulated as parkland and maintained by adjacent property owners. The District Department of Transportation (DDOT) maintains records for each street delineating the amount of public space within each right-of-way that is set aside for public parking. Parking width ranges from 2 ft to over 36 ft, although the most common width is 12 to 15 ft.

The most important regulations that maintain public parking's structure and appearance as parkland – collectively referred to as the Parking Regulations – are part of the District of Columbia Municipal Regulations, specifically *Title 24 Public Space and Safety*, reinforced with guidelines and standards in DDOT's *Design and Engineering Manual* (*32*) and policies in the Office of Planning's (OP) Comprehensive Plan. *DCMR Title 24* still defines public parking as, "that area of public space devoted to open space, greenery, parks, or parking that lies between the property line, which may or may not coincide with the building restriction line, and the edge of the actual or planned sidewalk that is nearer to the property line, as the property line and sidewalk are shown on the records of the District" (*33*). Parking regulations for fences, retaining walls, hedges, grading, and paving all reinforce its parklike character and uniform appearance. They also limit or preclude uses that detract from this character, such as drying laundry or, notably, storing vehicles.

The Public Space Committee continues to review public space applications required by *DCMR Title 24*, those seeking waivers from public space requirements, and additional applications the Committee chair determines need Committee review. Current District agencies represented in the Committee include DDOT, OP, Department of Consumer and Regulatory Affairs, and the Office of the Secretary. More recently, the Committee has played a key role in educating the general public about the importance and value of public parking as part of the city's park system and has the responsibility of determining when waivers from the parking regulations are appropriate.

Although public oversight has allowed a consistency and cohesion for a public parking system that is attractive and inviting to all District residents for almost 150 years, general knowledge of it and its benefits are not widespread. Many residents are surprised to learn what they consider to be their front yard is public space or that there is a building restriction line on their property that regulates an area as if it is in public space (*34*). Could renewed attention to the term "parking" help Washington and other cities revalue public space?

To that end, the District of Columbia has a renewed interest in the parking system and promoting it as a remarkable public resource. In addition to educating government agencies, property owners, and developers about public parking, the city has recently used the tradition of using public space as part of its parks and open space system as a way to bring passive recreation space to neighborhoods that lack traditional parks. Beginning in 2006, DDOT and OP developed streetscape guidelines for some areas of the city that reinterpret the meaning of public parking in a way that is more contemporary and addresses 21st century needs and challenges. For example, the Mount Vernon Triangle neighborhood's 55-ft wide sidewalks on K Street NW have been transformed into a linear park with lush landscaping and places for outdoor seating, sidewalk cafes, and over 5 acres of passive parkland. Along 1st Street NE in NoMa, the public parking is used to capture stormwater runoff from the roadway.

Implications for Public Space

Public space must continue to serve functional needs while maintaining an urban form for which the city is known. The magnitude of the District's right-of-way and open spaces is significant and represents about 45% of the city's land use—a major portion of the city's overall land area (*35*). The scope of its uses and impact on public health, safety, city character, and civic engagement is profound. The space serves the city's transportation network as well as a design feature that defines individual properties and the city as a whole. It is a shared commodity given context in DDOT's *Public Realm Design Manual*, which describes the city's wide avenues and long vistas as a defining characteristic of residential and commercial neighborhoods in the District and differentiate the city from surrounding jurisdictions (*36*). This image is the result of policies and regulations that few people are aware of and have established the unique character of the public parking (*37*).

Design of public space incorporating green space has broad value beyond aesthetics and also direct correlations with public health. The Center for Active Design attributes lack of spaces designed for activity, such as walking, as a core cause of obesity (*38*). Other studies have found a direct relationship between adding green space and reductions in crime (*39*). Studies by Jan Gehl understand street psychology in a way that delimits more visually complex, varied facades with openings and a multitude of uses as more attractive to people and which bring improvements to mood. This encourages people to pause and take in the view or at least slow down their gait (*39*). Such recognition has led to calls for investments in public space design that sound a lot like the District's parking system. Strategies included in the Urban Landscape Institute's Building Healthy Places Toolkit recommend pedestrian-oriented streetscapes and more access to nature by planting trees along sidewalks that encourage walking and provide better air quality, shade, and comfort (*40*).

Beyond designing for improved public space and health, contemporary urban designers have recognized the importance of a clear lexicon when designing streets. Andres Duany would call the street design the District has championed an example of "avenue" design, with adjacent landscaping and terminated vistas (*41*). John Stilgoe summarizes that, "Putting names to core components of landscape … empowers" (*42*). The District now has the opportunity to reemphasize a core aspect of its avenues, perhaps with idiolect, empowering today's designers of this space. When viewed in light of its unique public parking system and the documented benefits of green space, the tremendous value of this shared park system in the District becomes clear.

In nearly all cities, a majority of their land is devoted to public space and in some, upwards of 50% are paved streets, parking for cars, and sidewalks (*43*). This is typically designed to meet the requirements of a transportation network, but its design can achieve much more. The value of these spaces has been underestimated and there is an opportunity to develop portions of streets as parking intended for people. The studies described suggest a resulting benefit to health, well-being, and safety for our cities and neighborhoods. Other cities could consider retrofitting their public space to distribute parking along their roadways. This could be done in conjunction with overall efforts to bring complete streets (*44*) to transportation networks, explicitly prescribing a public parking area as a necessary element for making a corridor complete. Consideration for the overall public space comprising the transportation network should thus fundamentally include full design of the overall space between buildings, including the roadway, sidewalk, and public parking areas, and all these components should be considered as transportation projects are planned, designed, and implemented.

Conclusion

Thoughts for Today and the Future

Washington, D.C.'s unique public parking system creates a public realm that is both functional and beautiful. The original L'Enfant Plan, Parking Acts, and subsequent Highway and McMillan Plans heavily influenced how parking was implemented and resulted in a defining feature of the city. Limited contemporary knowledge around the origins of parking and its potential benefits leaves open the opportunity to reemphasize this special aspect of the built environment for the benefit of residents, workers, and visitors alike. Parking in the District benefits public health, safety, and the environment and provides continued opportunity to do justice to the intentions of Washington's planners.

The history of parking as a place to store artillery or as space for trees and other plants took a sharp turn by the end of the First World War. As automobile ownership increased, the shift in the meaning of parking for cars deeply affected one of the District of Columbia's foundational aspects. As parking became the place for cars, the knowledge of its history as green space for people was lost. The significance of this loss, coupled with the importance of public space for health, safety, and social benefits, argues for revival, as opposed to redefinition, of this feature (*45*). This paper, therefore, traces two significant topics: evolving ideas for how to use the public right-of-way and the changing vocabulary used to describe its components. This evolution points to opportunities for reimagining or restoring the future of urban street design by reflecting on the best of past precedents.

As guidance shifts rapidly toward better public transit infrastructure, complete streets, and walkable cities, it is time to rethink what the "parking" amenity is (*28*). Lessons from the District of Columbia's public parking system offer opportunities for communities around the world to consider similar approaches to preserve or add green space as they plan a transportation network to increase livability within their community. The composition and benefits of green space could be applied in other contexts, in other cities, and ultimately achieve parking in its truest sense: for people.

Acknowledgments

The authors appreciate the work completed in 2009 by Kale Anderson and Eli Pousson at the Office of Planning, documenting the history of parking regulations in the District of Columbia. The authors thank Steve Kehoe, formerly of DDOT, who provided additional insights in a 2017 interview.

Author Contributions

The authors confirm contribution to the paper as follows: study conception and design: RW, CS, RS; data collection: RW, CS, RS; analysis and interpretation of results: RW, CS, RS; draft manuscript preparation: RW, CS, RS. All authors reviewed the results and approved the final version of the manuscript.

References

1. Miller, C. *Carscape: A Parking Handbook*. Irwin-Sweeney-Miller Foundation, Columbus, IN, 1988.
2. Childs, M. *Parking Spaces: A Design, Implementation, and Use Manual for Architects, Planners, and Engineers*. McGraw-Hill, New York, NY, 1999.
3. Etymology of Park. Online Etymology Dictionary. http://www.etymonline.com/index.php?term = park. Accessed July 30, 2017.
4. Eno, W. The Storage of Dead Vehicles on Roadways. *The Annals of the American Academy of Political and Social Science*, Vol. 116, 1924, pp. 169–174.
5. Savage, K. *Monument Wars: Washington, D.C., The National Mall, and the Transformation of the Memorial Landscape*. University of California Press, Berkeley and Los Angeles, CA, 2009.
6. Tindall, W. The Origin of the Parking System of This City. *Records of the Columbia Historical Society*, Vol. 4, 1901, pp. 75–99.
7. Jakle, J. A., and K. A. Sculle. *Lots of Parking: Land Use in a Car Culture*. University of Virginia Press, Charlottesville, VA, 2004.
8. Lewis, S. *Main Street*. Chapter 3, Section 5. Harcourt, Brace & Co., New York, NY, 1920.
9. L'Enfant Plan. Library of Congress, Geography and Maps Division, 1792.
10. Anderson, K., and E. Pousson. *Historic Context Statement on Public Space in the District of Columbia*. District of Columbia Office of Planning, Washington, D.C., 2009.

11. Scott, P. "The City of Living Green": An Introduction to Washington's Street Trees. *Washington History*, Vol. 18, No. 1/2, 2006, pp. 20–45.
12. Willard, H. K. Henry Augustus Willard: His Life and Times. *Records of the Columbia Historical Society*, Vol. 20, 1917, pp. 270–272.
13. U.S. Congress. An Act Authorizing the Corporation of the City of Washington to Set Apart Portions of Streets and Avenues as Parks for Trees and Walks. Presented at 41st Congress, 2nd Session, Washington, D.C., 1870.
14. U.S. Congress. An Act for the Protection of Parks in Streets and Avenues. Presented at 2nd Legislative Assembly, 1st Session, Washington, D.C., 1872.
15. *Highway Act Historic Context Report*. EHT Traceries. 2009.
16. U.S. Congress. An Act for the Protection of Parks and Lawns in the District of Columbia, and reservations and other land of the United States in said District. Presented at 3rd Legislative Assembly, 1st Session, Washington, D.C., 1873.
17. U.S. Congress. An Act to Provide for the Establishment of Building Lines on Certain Streets in the District of Columbia, and for Other Purposes. Presented at 56th Congress, 3rd Session, Washington, D.C., 1899.
18. Davis, T. *Designing the Nation's Capital: The 1901 Plan for Washington, D.C.* U.S. Commission of Fine Arts, Washington, D.C., 2006.
19. Shaheen, C. M. *Beyond the Grand Design: City Planning in Washington beyond the Federal Core, 1919-1941.* Master's thesis. The George Washington University, Washington, D.C., 2000.
20. Board of Public Works. *Evening Star*. July 16, 1873, p. 2.
21. The Funeral Procession. *Evening Star*. November 26, 1875.
22. Obstructed Streets. *Evening Star*. December 9, 1876.
23. Too Many Cars are Parked on the Virginia Avenue Tracks. *Evening Star*. October 9, 1895, p. 2, and "Reservation No. 266", June 15, 1897.
24. Shoup, D. *The High Cost of Free Parking*. APA Planners Press, Chicago, IL, 2011.
25. Biodiversity Heritage Library. https://www.biodiversitylibrary.org. Accessed July 30, 2017.
26. Maloney, D. *The Governmental Response to the Parking Problem in Washington, D.C. 1920-1980*. Graduate School Paper. The George Washington University, Urban and Regional Planning 208, Washington, D.C., 1980.
27. Report of the Board of Commissioners to Congress, 1924. Board of Commissioners of the District of Columbia.
28. Richmond, M. The Etymology of Parking. *Arnoldia*, Vol. 73, No. 2, 2015, pp. 19–24.
29. Hackman, L., and N. D. Martin. *The Parking Industry: Private Enterprise for the Public Good.* National Parking Association, Washington, D.C., 1969.
30. Norton, P. *Fighting Traffic: The Dawn of the Motor Age in the American City*. The MIT Press, Cambridge, MA, 2008.
31. Vaux, B., and S. Golder. *The Harvard Dialect Survey*. Harvard University Linguistics Department, Cambridge, MA, 2003. http://www4.uwm.edu/FLL/linguistics/dialect/staticmaps/q_60.html.
32. District Department of Transportation. *Design and Engineering Manual*. District Department of Transportation, Washington, D.C., 2017.
33. District of Columbia Municipal Regulations. *Title 24: Public Space and Safety*. https://www.dcregs.dc.gov/Common/DCMR/RuleList.aspx?ChapterNum=24-1. Accessed July 30, 2017.
34. Shaheen, C. Public Parking: Transforming the Nation's Capital from Gritty to Garden City. Presented at 38th Annual Conference on DC Historical Studies, Washington, D.C., 2011.
35. District of Columbia Office of Planning. *The Comprehensive Plan*. 2011.
36. District Department of Transportation. *Public Realm Design Manual: A Summary of District of Columbia Regulations and Specification for the Design of Public Space Elements*. District Department of Transportation, Washington, D.C., 2011.
37. American Planning Association. *Characteristics and Guidelines of Great Public Spaces*. https://www.planning.org/greatplaces/spaces/characteristics.htm/. Accessed July 30, 2017.
38. Center for Active Design. *Design + Health*. https://centerforactivedesign.org/data/. Accessed July 31, 2017.
39. Montgomery, C. *Happy City: Transforming Our Lives through Urban Design*. Farrar, Straus, and Giroux, New York, NY, 2013.
40. Urban Land Institute. *Building Healthy Places Toolkit: Strategies for Enhancing Health in the Built Environment*. Urban Land Institute, Washington, D.C., 2015.
41. Duany, A. *The Lexicon of New Urbanism*. DPZ, Miami, FL, 2003.
42. Stilgoe, J. *What is Landscape?* The MIT Press, Cambridge, MA, 2015.
43. Litman, T. *Pavement Busters Guide: Why and How to Reduce the Amount of Land Paved for Roads and Parking Facilities*. Victoria Transport Policy Institute, Victoria, BC, Canada, 2017.
44. Smart Growth America. *What are Complete Streets?* https://smartgrowthamerica.org/program/national-complete-streets-coalition/what-are-complete-streets/. Accessed November 12, 2017.
45. Alpert, D. *What Should We Call the "Parking" Area?* Greater Greater Washington, 2008. https://ggwash.org/view/836/what-should-we-call-the-parking-area. Accessed June 7, 2017.

The Standing Committee on Transportation History (ABG50) peer-reviewed this paper (18-06120).

Sage Publications USA
sagepub.com

ISBN 978-1-5108-7822-8

9 781510 878228

CURRAN ASSOCIATES INC.
proceedings.com

International Conference on Maritime Innovation and Emerging Technologies 2021

Online
17 - 18 March 2021

The Royal Institution of
Naval Architects